# Promoting Chemical Laboratory Safety and Security in Developing Countries



Committee on Promoting Safe and Secure Chemical Management in Developing Countries

Board on Chemical Sciences and Technology

Division on Earth and Life Studies

NATIONAL RESEARCH COUNCIL
*OF THE NATIONAL ACADEMIES*

THE NATIONAL ACADEMIES PRESS
Washington, D.C.
**www.nap.edu**

**THE NATIONAL ACADEMIES PRESS 500 Fifth Street, N.W. Washington, DC 20001**

NOTICE: The project that is the subject of this report was approved by the Governing Board of the National Research Council, whose members are drawn from the councils of the National Academy of Sciences, the National Academy of Engineering, and the Institute of Medicine. The members of the committee responsible for the report were chosen for their special competences and with regard for appropriate balance.

This study was funded under grant number S-LMAQM-08-CA-140 from the United States Department of State. The opinions, findings and conclusions stated herein are those of the authors and do not necessarily reflect those of the United States Department of State.

International Standard Book Number-13: 978-0-309-15041-5
International Standard Book Number-10: 0-309-15041-8

Additional copies of this report are available from the National Academies Press, 500 Fifth Street NW, Lockbox 285, Washington, DC 20055; (800) 624-6242 or (202) 334-3313 (in the Washington metropolitan area); Internet, *http://www.nap.edu.*



Printed in the United States of America

# THE NATIONAL ACADEMIES

***Advisers to the Nation on Science, Engineering, and Medicine***

The **National Academy of Sciences** is a private, nonprofit, self-perpetuating society of distinguished scholars engaged in scientific and engineering research, dedicated to the furtherance of science and technology and to their use for the general welfare. Upon the authority of the charter granted to it by the Congress in 1863, the Academy has a mandate that requires it to advise the federal government on scientific and technical matters. Dr. Ralph J. Cicerone is president of the National Academy of Sciences.

The **National Academy of Engineering** was established in 1964, under the charter of the National Academy of Sciences, as a parallel organization of outstanding engineers. It is autonomous in its administration and in the selection of its members, sharing with the National Academy of Sciences the responsibility for advising the federal government. The National Academy of Engineering also sponsors engineering programs aimed at meeting national needs, encourages education and research, and recognizes the superior achievements of engineers. Dr. Charles M. Vest is president of the National Academy of Engineering.

The **Institute of Medicine** was established in 1970 by the National Academy of Sciences to secure the services of eminent members of appropriate professions in the examination of policy matters pertaining to the health of the public. The Institute acts under the responsibility given to the National Academy of Sciences by its congressional charter to be an adviser to the federal government and, upon its own initiative, to identify issues of medical care, research, and education. Dr. Harvey V. Fineberg is president of the Institute of Medicine.

The **National Research Council** was organized by the National Academy of Sciences in 1916 to associate the broad community of science and technology with the Academy's purposes of furthering knowledge and advising the federal government. Functioning in accordance with general policies determined by the Academy, the Council has become the principal operating agency of both the National Academy of Sciences and the National Academy of Engineering in providing services to the government, the public, and the scientific and engineering communities. The Council is administered jointly by both Academies and the Institute of Medicine. Dr. Ralph J. Cicerone and Dr. Charles M. Vest are chair and vice chair, respectively, of the National Research Council.

**www.national-academies.org**

## COMMITTEE ON PROMOTING SAFE AND SECURE CHEMICAL MANAGEMENT IN DEVELOPING COUNTRIES

**Ned D. Heindel** (*Chair)* Lehigh University, Bethlehem, PA
**Charles Barton**, Independent Consultant, San Ramone, CA
**Janet S. Baum**, Independent Consultant, University City, MO
**Apurba Bhattacharya**, Texas A&M University, Kingsville
**Charles P. Casey**, University of Wisconsin, Madison
**Mark C. Cesa**, INEOS USA, LLC, Naperville, IL
**M. Iqbal Choudhary**, University of Karachi, Pakistan
**Robert H. Hill**, Battelle Memorial Institute, Atlanta, GA
**Robin M. Izzo**, Princeton University, NJ
**Russell W. Phifer**, WC Environmental, LLC, West Chester, PA
**Mildred Z. Solomon**, Harvard Medical School, Boston, MA
**James M. Solyst**, ENVIRON, Arlington, VA
**Usha Wright**, O'Brien & Gere, Syracuse, NY
**Patrick J. Y. Lim**, University of San Carlos, Cebu City, Philippines

Staff

**Dorothy Zolandz,** Director, Board on Chemical Sciences and Technology
**Kathryn Hughes,** Program Officer
**Tina M. Masciangioli,** Responsible Staff Officer
**Sheena Siddiqui,** Research Assistant
**Jessica Pullen**, Administrative Coordinator, through May 2010
**Lynelle Vidale,** Senior Program Assistant, through March 2010
**Norman Grossblatt,** Senior Editor

## BOARD ON CHEMICAL SCIENCES AND TECHNOLOGY

# Acknowledgment of Reviewers

This report has been reviewed in draft form by persons chosen for their diverse perspectives and technical expertise in accordance with procedures approved by the National Research Council's Report Review Committee. The purpose of this independent review is to provide candid and critical comments that will assist the institution in making the published report as sound as possible and to ensure that it meets institutional standards of objectivity, evidence, and responsiveness to the study charge. The review comments and draft manuscript remain confidential to protect the integrity of the deliberative process. We thank the following for their review of the report:

**Asad Abidi**, University of California, Los Angeles
**Mukund Chorghade**, Chorghade Enterprises, Natick, Massachusetts
**Peter Dorhout**, Colorado State University, Fort Collins
**Kenneth Fivizzani**, Independent Consultant, Naperville, Illinois
**Alastair Hay**, University of Leeds, United Kingdom
**Claude Lucchesi**, Northwestern University, Evanston, Illinois
**Richard Niemeier**, National Institute for Occupational Safety and Health, Cincinnati, Ohio
**Khalil Qureshi**, Lahore University of Management Sciences, Pakistan
**Elsa Reichmanis**, Georgia Institute of Technology, Atlanta
**Carolyn Ribes**, Dow Benelux B.V., Terneuzen, Belgium

Although the reviewers listed above provided many constructive comments and suggestions, they were not asked to endorse the conclusions or recommendations, nor did they see the final draft of the report before its release. The review of the report was overseen by **R. Stephen Berry,** University of Chicago, Illinois, and **Jeffrey I. Steinfeld,** Massachusetts Institute of Technology, Cambridge. Appointed by the National Research Council, they were responsible for making certain that an independent examination of the report was carried out in accordance with institutional procedures and that all review comments were carefully considered. Responsibility for the final content of the report rests entirely with the authors and the institution.

# About This Study

There is growing concern about the possible use of toxic industrial chemicals or other hazardous chemicals by those seeking to perpetrate acts of terrorism. The U.S. Chemical Security Engagement Program (CSP), funded by the U.S. Department of State and run by Sandia National Laboratories, seeks to develop and facilitate cooperative international activities that promote best practices in chemical security and safe management of toxic chemicals, including

- partnering with host governments, chemical professionals, and industry to assess and fill gaps in chemical security abroad;
- providing technical expertise and training to improve best practices in security and safety among chemical professionals and industry;
- increasing transparency and accountability for dangerous chemical materials, expertise, and technologies; and
- providing opportunities for collaboration with the international professional chemical community.

The Department of State called on the National Academies to assist in the CSP's efforts to promote chemical safety and security in developing countries.[1] More specifically, the National Research Council was asked to perform two main tasks focused on laboratory-scale activities.

[1]See Appendix A for the full statement of task.

1. Examine the dual-use risks posed by toxic industrial chemicals and other hazardous chemicals in developing countries, particularly in regions where terrorism is on the rise.

2. Provide guidance and produce educational materials on a baseline of practices in the handling and storage of hazardous chemicals required to promote safety and security in their use in the developing world.

This report addresses these two tasks: the educational materials described will be delivered to the sponsors separately from this report. The educational materials will be based on material generated by this current study (mainly Chapters 3 and 4) and derived from the forthcoming revised edition of *Prudent Practices in the Laboratory: Handling and Management of Chemical Hazards* (The National Academies Press, Washington, D.C., 2010). Those materials will be produced upon completion of the current report, and are meant to be used by CSP and similar organizations that are engaged in chemical laboratory safety and security outreach in developing countries. The NRC plans to seek assistance from other organizations such as the International Union of Pure and Applied Chemistry, the Organization for the Prohibition of Chemical Weapons, and the Academy of Sciences for the Developing World (TWAS) to review, co-brand, and distribute the materials. The materials will be distributed in print and electronic format, and will be translated into Arabic, French, and Indonesian.

Appointed by the National Research Council, the Committee on Promoting Safe and Secure Chemical Management in Developing Countries was convened to carry out the task. The 14 expert committee members represent the fields of chemical safety, chemical security, chemical management, environmental health and safety, international chemical exchanges and scientific affairs, organic and pharmaceutical chemistry, industrial hygiene and safety, biological safety and security, toxicology, laboratory design and safety, education and behavioral change, and basic chemical synthesis. They have experience in industrial and academic laboratory-scale activities, and they include researchers who have firsthand experience with the conduct of chemical research in developing countries. The committee held four meetings, three of which included data-gathering sessions.[2]

The committee would especially like to acknowledge the two international members of the committee, **Iqbal Choudhary** of Pakistan and **Patrick Lim** of the Philippines, who traveled very long distances and crossed many time zones to attend the committee meetings. It also thanks the three international guest speakers who made long journeys from developing countries to one of our data-gathering meetings and provided invaluable insight to the committee:

[2]See Appendix C for committee member and guest speaker biographical information.

**Supawan Tantayanon**, Chulalongkorn University, Bangkok, Thailand
**Engida Temechegn**, Addis Ababa University, Ethiopia
**Khalid Temsamani**, University Abdelmalek Essaadi, Tétouan, Morocco

In addition, we thank **Mohammad El-Khateeb** of Jordan University of Science and Technology, who joined one meeting via video conference; and **Alastair Hay** of the University of Leeds, United Kingdom, and **Richard W. Niemeier** of the U.S. National Institute for Occupational Safety and Health, Cincinnati, Ohio, who had shorter journeys but provided no less valuable insights to the committee.

The committee hopes this report will serve the needs of the Department of State and the CSP and the chemical safety and security needs of the larger international chemistry community.

The Committee on Promoting Safe and Secure
Chemical Management in Developing Countries

# Abbreviations

| | |
|---|---|
| ACC | American Chemistry Council |
| ACS | American Chemical Society |
| AIHA | American Industrial Hygiene Association |
| ASSE | American Society of Safety Engineers |
| BCSP | Board of Certified Safety Professionals |
| CCS | ACS Committee on Chemical Safety |
| CHAS | ACS Division of Chemical Health and Safety |
| CHO | Certified Chemical Hygiene Officer |
| CIH | Certified Industrial Hygienist |
| COC | Chemical of Concern |
| CSHEMA | Campus Safety Health and Environmental Management Association |
| CSJ | Chemical Society of Japan |
| CSO | Chemical Safety Officer |
| CSP | Chemical Security Engagement Program |
| CWC | Chemical Weapons Convention |
| FACS | Federation of Asian Chemical Societies |
| FASC | Federation of African Societies of Chemistry |
| IAC | ACS Committee on International Activities |
| ICCA | International Council of Chemical Associations |
| IPCS | International Program on Chemical Safety |
| IUPAC | International Union of Pure and Applied Chemistry |
| IYC 2011 | International Year of Chemistry 2011 |

| | |
|---|---|
| MSDS | Materials Safety Data Sheet |
| NAO | National adhering organizations |
| NCGC | National Core Group in Chemistry |
| NIOSH | National Institute for Occupational Safety and Health |
| OECD | Organisation for Economic Co-operation and Development |
| OPCW | Organization for the Prohibition of Chemical Weapons |
| PacifiChem | International Chemical Congress of Pacific Basin Societies |
| PPE | Personal protective equipment |
| RCSC | Responsible Care® Security Code |
| REACH | Registration, Evaluation, Authorization, and Restriction of Chemical Substances |
| SAICM | Strategic Approach to International Chemicals Management |
| SME | Small and Medium-sized Enterprises |
| SOP | Safe Operating Procedure |
| SVA | Security Vulnerability Assessment |
| UNESCO | U.N. Educational, Scientific, and Cultural Organization |

# Contents

Appendixes

# Summary and Recommendations

International chemical security has historically focused on chemicals controlled by the Chemical Weapons Convention (CWC),[1] as well as explosives, flammables, and chemicals used for production of illicit drugs. In the last 20 years, however, there has been growing concern about the use of toxic industrial chemicals and other hazardous chemicals by those seeking to perpetrate acts of terrorism[2]; such chemicals are commonly known as dual-use or multiple-use chemicals. The 1995 sarin attack by the Aum Shinrikyo cult brought to light the potentially devastating outcome of the diversion and use of chemicals for malevolent purposes. Many other incidents have occurred involving intentional releases of hazardous chemicals with significance to public health and safety.[3]

In this report, the term chemicals of concern (COC) is used to describe all those laboratory chemicals that pose a high risk to safety and security and include[4]

---

[1]See Appendix Table D-1 for an example list of Chemical Weapons Convention chemicals, and see the U.S. Chemical Weapons Convention Web site at *www.cwc.gov* for further information.

[2]For more information, see Hauschild, V.D., and G.M. Bratt. Prioritizing industrial chemical hazards. *Journal of Toxicology and Environmental Health, Part A* 68(2005):857:876.

[3]For an excellent review, see M.M. Patel, J.G. Schier, and M.G. Belson. Recognition of illness associated with covert chemical releases. *Pediatric Emergency Care* 22(2006):592-601.

[4]See Appendix D for a sample list of chemicals of concern.

1. chemicals likely targeted for theft or diversion
   a. CWC chemicals;
   b. explosives and improvised explosive device precursors; and
   c. mass Effect Agents and Precursors.
2. chemicals with high acute toxicity (Globally Harmonized System Category 1); and
3. chemicals used in clandestine production of illicit drugs.

At the same time, laboratory chemists throughout the world work daily with many potentially hazardous chemicals, including COCs, for legitimate purposes and generally follow the necessary safety procedures for handling and disposal of these chemicals. Chemical laboratories are where chemical research, development, and education take place. Chemical manufacturers also use laboratories for quality control, process monitoring, and analysis related to compliance with government regulations. The quantities of chemicals used in such settings are typically small and pose less risk compared with industrial-scale manufacturing, use, and transport of chemicals. Chemical laboratories in small-scale industrial and academic settings, however, tend to operate independently, have less government and regulatory oversight, and are generally more accessible to the public than large-scale industrial laboratory and manufacturing facilities. Such laboratories thus present a vulnerable target for those seeking to do harm. For example, in 2002 Joseph Konopka (a.k.a. "Dr. Chaos") was found to be storing over a pound of cyanide compounds and other hazardous chemicals in a tunnel near the Chicago subway system, and at least part of the stores of cyanide were obtained from the campus of the University of Illinois at Chicago.[5]

The growing security threat of COCs thus presents a new challenge to working with chemicals in the laboratory, especially in small-scale industrial and academic settings. While large-scale industrial manufacturing and use of COCs is a dominant concern in national and international chemical security, use of chemicals at the laboratory scale poses a unique and significant security threat and is the main focus of this report.

Developing countries in particular face many challenges with regard to chemical laboratory safety and security. They are generally characterized as having low- to lower-middle-income economic status,[6] but they can vary widely in socioeconomic standards and implementation of the rule of law. Some first-rate institutions in developing countries have excellent labora-

[5]United States v. Joseph Konopka, U.S. District Court, Criminal Complaint Case Number 02 CR, March 9, 2002, Cook County, Northern District of Illinois, Eastern Division; and CNN.com, March 12, 2002. Man allegedly stored cyanide in Chicago subway. *http://archives.cnn.com/2002/US/03/12/chicago.cyanide/index.html* (accessed December 17, 2009).

[6]See the World Bank country classifications: *http://data.worldbank.org/about/country-classifications* (accessed May 6, 2010).

tory safety and security systems in place, and some national governments in developing countries have established policies for occupational safety in the work place.[7] However, laboratory safety and security are generally not a high priority in developing countries. In addition, developing countries that do have a legal framework of laws and regulations for chemical safety often lack an adequate and effective system of enforcement. As developing countries become more economically competitive and strive to increase chemistry activity, they face many challenges in improving laboratory safety and security. Safety and security practices are intended to help laboratories carry out their primary functions efficiently, safely, and securely, but improving safety and security is often seen as inhibitory rather than enabling.

## CURRENT SAFETY AND SECURITY PRACTICES IN DEVELOPING COUNTRIES

> The culture of laboratory safety depends ultimately on the working habits of individual chemists and their sense of teamwork for protection of themselves, their neighbors, and the wider community and environment. . . . Safety in the laboratory also depends on well-developed administrative structures and supports that extend beyond the laboratory's walls within the institution.
>
> National Research Council. 2005. *Prudent Practices in the Laboratory: Handling and Disposal of Chemicals,* 1995 (Washington, D.C.: National Academies Press).

This section provides an overview of current chemical laboratory safety and security practices in developing countries, largely focused on the barriers to and needs for improvement. The information presented is based on the collective experience of the committee members and the insightful guest speakers listed in "About This Study," including several from representative developing countries.

### Current Safety Practices

Chemical laboratories in developing countries have large numbers of students in teaching laboratories, but they typically have a relatively small (although increasing) number of people engaged in high-level research. In general, use of hazardous laboratory chemicals is greater in institutions that offer graduate programs and that engage in basic research; but that is

[7]For example, see country policies listed on the International Labor Organization Web site: *http://www.ilo.org/public/english/region/asro/bangkok/asiaosh/std_leg/national/indexnat.htm* (accessed December 18, 2009).

a generalization, and there is a wide variation in such activities and safety practices within and between countries.

The increasingly global interconnectivity of science, driven by the proliferation of mobile phones, air travel, e-mail, and the Internet, has resulted in scientists everywhere becoming more aware of laboratory best practices and in some cases the prestige that attends recognition by the international community. This has led some institutions in developing countries to seek to attain certifications in international standards[8] (such as ISO 9001, Quality Management Systems standard; ISO 14001, Environmental Management Systems standard; and ISO 17025, general requirements for the competence of testing and calibration laboratories), which has played an important role in including occupational and community safety as a component of the overall standards system.

The barriers to and needs for improving laboratory safety practices in developing countries are listed here and described in detail below.

**Barriers**
Financial limitations
Climate constraints
Cultural challenges

**Needs**
Institutional safety policy and rules
Institutional implementation strategies or plans
General safety awareness and training
Reporting and compliance processes
Waste disposal systems
External help and support

## Barriers to Improving Safety Practices

### Financial Limitations

Financial constraints are among the most important bottlenecks in implementing safety practices in chemical laboratories in developing countries. They affect every aspect of safety plans and implementation, because initial investments and sustained support are required to build and maintain a safety infrastructure.

Laboratory buildings are specialized structures, and the addition of safety features increases the costs of planning and construction. In some

[8]See International Organization of Standardization at *http://www.iso.org/iso/home.htm* (accessed December 18, 2009).

cases the extra cost may be quite small (2 percent to 3 percent of total capital cost of building the laboratory), but in an effort to save money, new laboratory buildings are often constructed with inadequate safety provisions. Some chemical laboratories are situated on upper floors of high-rise buildings in highly populated urban areas and have no provision for an exit plan or separate chemical stores. The highest parts of some of those buildings are beyond the reach of firefighting ladders. In general, there is little consultation between the chemists, who are the end users of the laboratories, and the architects, builders, and chemical safety experts.

Similarly, a sufficient supply of operating fume hoods, fire extinguishers, and other protective equipment requires funding, which is often unavailable. In many teaching laboratories, a large number of students are assigned to work in a single chemical-fume hood, and this makes such equipment largely useless.

Financial constraints are largely responsible for unfavorable student-to-teacher ratios in many teaching laboratories (for example, 40:1 in Ethiopia and 25:1 in the Philippines), which not only affect the quality of teaching but also make laboratory safety challenging. Hiring, training, and retaining of safety personnel is difficult in such a financially constrained environment. Many laboratory staff who have attained a high level of proficiency and competence leave academic laboratories for more lucrative positions in industry or even employment abroad. Other laboratory staff may feel they have little choice but to tolerate unsafe jobs because of financial constraints. They are forced to choose between keeping a job and being safe.

## Climate Constraints

Special climatic conditions in many developing countries hinder compliance with safety practices. Many regions of the world experience extremes in weather and have no provision for controlling indoor temperature or humidity other than with the use of ceiling fans and windows. Students in the hot and humid environments of tropical and subtropical regions often do not wear chemical splash goggles or latex gloves because they are uncomfortable. While institutions tend to schedule closures or vacations during extreme weather, appropriate provisions cannot always be made for storing chemicals safely during such conditions.

## Cultural Challenges

Differences in culture have a substantial effect on behavior, including chemical safety and security. Developing countries often have a hierarchical structure in which decisions are made and implemented from the top down. In such a management structure, a large commitment from leadership, which is a component in scientific methods and practice, is required before any progress can be made.

However, many of those in leadership positions take responsibility without being held accountable. Such a culture can discourage recognition of safe behavior and prevent criticism of unsafe or suspicious conduct by superiors or even peers. Hierarchical management structures thus can inhibit reliance on coworkers to report or prevent breaches in safety or security.

## Needs for Improving Chemical Laboratory Safety Practices

### Institutional Safety Policy and Rules

Not many institutions in the developing world have specific safety rules and policies. They generally rely on generic safety practices that may not be very clear or known to people who are supposed to follow and implement them. Even where safety policies exist, safety in developing countries tends to depend on personal initiative, and safety practices often deteriorate when a strong advocate for safety is promoted, retires, or loses authority.

In some cases, government regulations are targeted at the chemical or manufacturing industries, and many of them are concerned primarily with waste management. However, government agencies tasked to institute and implement the regulations often lack the resources and trained enforcement staff needed to be effective. Most agencies can barely police industry, let alone private and academic laboratories. In addition, the regulations appropriate for large-scale industrial operations are not readily adaptable to academic laboratories.

### Institutional Implementation Strategies or Plans

In many developing countries where rules and regulations exist, universities and R&D institutions often fail to implement them. Part of the failure stems from institutions copying policies verbatim from more developed countries without seeking input from end users who work in the local environments. Working relations are also often poor between the government enforcing agencies and private or state-run institutions. The planning and implementation of safety rules demand a strong sense of responsibility, and long-term commitment from leaders. Academic leaders in developing countries either are not aware of the importance of safety in the workplace or do not have the means to implement safety rules fully. As discussed earlier, academic leaders struggle to get essential funding to run their institutions; after they have paid for salaries and supplies, little funding is available to provide the infrastructure and human resources needed to implement institutional safety plans.

### General Safety Awareness and Trained Safety Personnel

Safety is not on the mainstream academic agenda in many developing countries. It is hardly discussed in meetings of faculty members or managers of chemical laboratories, and there are minimal safety instruction, teaching, or training workshops for safety offices. Faculty and laboratory managers who demonstrate consistently safe behavior typically go unnoticed. Furthermore, aside from a perfunctory orientation, there is almost no formal safety training of students or staff members before they are allowed to work in chemical laboratories. Laboratory safety is not a part of the regular teaching curriculum, except in a few universities. It is more common to find training in content and pedagogy rather than safety. Notably, in many chemistry conferences and congresses throughout the world, not just in developing countries, there is little time devoted to safety topics. As a result, there is a serious lack of appreciation for safe practices.

At the faculty level, the notion of academic freedom is often misused to avoid compliance with safety regulations, and faculty members typically cannot be forced to comply with safety rules. Although every chemistry department has faculty members who were educated and trained in Western universities that have higher safety standards, the effect of their training is barely felt. Because safety has very low priority, there is hardly any safety instruction, and there are few training workshops for safety officers for this purpose.

### Reporting Systems

Reporting of incidents (such as chemical spills, fires, and missing supplies) is one of the most difficult components to implement in any safety system. In developing countries, cultural barriers and fear of punitive action, generally lead to failure of reporting incidents. This in turn results in missed opportunities for lessons learned and continuous improvement in safety. Further, the low numbers that are reported give an unwarranted impression that there are no safety issues. The lack of reporting is commonly based on inappropriate definitions. Incidents are "minor" if they do not involve major burns or loss of an organ, or a life, and many are not reported. There is no concept of reporting "near escape" incidents (commonly called "near misses"); it is as though the incidents had never happened, so there is no learning that can be shared. However, this is a pervasive problem throughout the world, not just in developing countries.

### Waste Disposal Systems

Many developing countries have no proper waste disposal facilities or systems, or do not know how to implement waste disposal cost-effectively, especially for laboratory chemicals. Organic solvents are disposed of by being poured into the municipal drain system, and solid wastes are generally dumped into garbage spaces or burned in open spaces. Some countries prohibit incinerating organic solvents, but do not provide alternative disposal systems, treatment of chemical waste onsite is rare, and there are no specialized waste disposal companies.

In a few institutions, however, waste water from chemical laboratories is collected and treated in campus facilities and recycled for irrigation. Where national regulatory authority exists for working with radioisotopes, government agencies typically collect and dispose of radioactive waste from laboratories. Most developing countries do not have similar arrangements for disposal of other hazardous chemicals.

### External Help and Support

In most developing countries, external help is not always available when it is needed. Fire departments and ambulances may come on call, but they do not always have adequate manpower and equipment or the proper training to handle hazardous chemical emergencies. Further complications arise due to lack of proper emergency preparedness in coordination with academic institutions, especially for specialized situations involving hazardous chemicals. Therefore, institutions are often left to make their own arrangements to handle or not handle emergencies. In some instances universities have their own fire trucks, ambulances, and medical emergency centers, but this is uncommon.

Similarly, regulatory agencies equivalent to the U.S. Environmental Protection Agency or U.S. Occupational Safety and Health Administration, where they exist, may not have jurisdiction over laboratory-scale operations, or they are ineffective. As a result, there is no external audit of safety practices of chemical laboratories.

## Current Security Practices

On a positive note, most campuses in developing countries have strict entry policies, probably better than most campuses in developed countries. Students must possess valid identification, and vehicles are screened upon entry and exit. Security guards are stationed at entry gates, and roving guards patrol the buildings. However, the purpose is largely to maintain law and order on highly politicized campuses or to prevent theft of equipment, not for security of chemical storage.

In teaching settings, access to chemical laboratories and storage areas is often controlled by specialized permits and follows strict protocols for working hours. In research settings, however, students sometimes work alone at night, often without emergency contact information.

Increased international communications using e-mail and the Internet, air travel, and telecommunications have also made scientists in all countries much more aware of international agreements, such as the Chemical Weapons Convention, and of best practices for the secure use and storage of chemicals, even on a laboratory scale. However, many chemists throughout the world still do not know about such security practices, and currently there are no certifications in international standards of security like those for safety and environmental protection.

Many laboratories in developing countries have basic secure storage of chemicals to prevent theft by outsiders. However, most are in need of procedures to ensure that there is no diversion by individuals working in laboratories or visitors, including friends and relatives. The barriers to and needs for improving chemical security practices in developing countries are listed here and described in detail below.

**Barriers**
Financial limitations
Bulk purchase of chemicals
Cultural challenges

**Needs**
Institutional security policies
Institutional security management plans, equipment, and services
General security awareness
Trained and motivated security personnel
Reporting systems

## Barriers to Improving Security Practices

### Financial Limitations

Secure storage of chemicals, especially COCs, requires stringent measures such as a separate building with lockable doors and an alarm system. Such infrastructure is nonexistent in most developing countries, and all chemicals, even organic solvents and peroxides, are generally stored in small storage areas in laboratory buildings or at laboratory benches. Gas cylinders are typically stored just outside laboratories, often on ledges that are exposed to the elements.

### Bulk Purchase of Chemicals

Most of the universities in the developing world are primarily teaching institutions. Very few universities have active research programs and research centers that require exotic and specialized chemicals. As a result, only common chemicals are purchased and stored, often in inadequate departmental facilities. Few universities have central storage facilities especially constructed for chemicals.

Common teaching laboratory chemicals are typically procured through open bidding at competitive prices in bulk quantities whenever funds are available. That creates a feast-or-famine situation in which it is often difficult to update inventory and ensure security. Inventory is usually maintained for the financial audit, not the benefit of professors or researchers. The concept of just-in-time use is common in highly industrialized nations but essentially unknown in developing countries. It is not unusual for procurement of a single chemical to take as long as six months. Thus, large quantities of bulk chemicals are routinely stored in academic institutions to prevent shortages while the arrival of an order is awaited. Both the transportation and storage of such large quantities presents a security challenge for any institution.

Inventory control is generally in the form of log books or computer-based inventory systems. Chemicals are issued through written requests on prescribed forms, but there is no tracking of chemicals after they are issued to individual laboratories. No record of their use and disposal is kept. Bar codes and radio frequency identification tagging are available in some developing countries but are not commonly used for keeping track of where or how a chemical is used. Chemicals procured by individual faculty members with project funds are often not recorded in the central inventory; faculty members are expected to add them to the central inventory voluntarily but have little incentive to do so. This is also true for laboratories in most developed countries.

Most inventory-tracking efforts are geared to the control of purchasing and storage of chemicals and reagents. There is little tracking of substances that are synthesized or isolated in research laboratories. For example, toxins and other bioactive substances may be obtained from natural terrestrial and marine organisms, but there is often no reporting of them until publication in a journal.

### Cultural Challenges

Differences in cultures in developing countries can also affect security practices in the laboratory. Cultural challenges include excessive hierarchy of power, too much trust in others, lack of a culture of sharing, lack of accountability, hiding or not reporting security lapses, frequent visits to the labs by friends and acquaintances, and not recognizing good security practices of others.

## Needs for Improving Laboratory Security Practices

### Institutional Security Policies

Chemical facility security is rarely treated separately from general security concerns. At both national and institutional levels, there are no plans for security of chemical facilities and storage sites. Vulnerability assessments are rare, and hidden dangers are poorly understood. In the absence of a specialized security strategy, generic security measures are applied to chemical laboratory facilities.

In some cases, strict regulations are in place for the procurement of chemicals that are identified as precursors in the manufacture of illicit drugs. One such example is acetone, a very common laboratory solvent. Academic institutions, suppliers, and distributors have to register and secure a license from a drug enforcement agency before they can purchase such chemicals. A similar setup is in place for explosive precursors, for which a license must be obtained from law enforcement agencies. There are, however, malpractices associated with licensing, such as use of an invalid license, obtaining a license through illegal means, or bribing inspectors.

Like safety, security depends on personal initiative in many developing countries. Once in a while an administrator who takes security seriously comes along, but security concerns return to being unattended when he or she moves out of the system.

### Institutional Security Management Plans, Equipment, and Services

Even when a security plan does exist in a developing country, the plan usually fails at the implementation stage. Strategies for implementing security plans are generally nonexistent. Concern about security rarely goes beyond preventing ordinary theft by outsiders. Generic security practices are usually implemented in developing countries; they often involve check points at gates, roving patrols, and the stationing of security guards in every building. Security aspects of management of chemicals that may serve as building blocks for extremely hazardous materials that can be used against the public have no priority. Part of the reason is a lack of information, and part is a lack of conviction that such use could occur in one's institution.

### General Security Awareness

The importance of security is recognized in developing countries, but the threat and danger arising from lack of security of chemical storage facilities is not properly understood. There is little knowledge regarding the security of COCs on campuses in developing countries. Chemists may be aware of the potential of dual-use chemicals, but top administrators have little or no knowledge of the dangers posed by such chemicals.

### Trained and Motivated Security Personnel

Proper security of chemical storage and facilities, especially those with COCs, requires not only proper infrastructure but also adequately trained and motivated staff. In most developing countries, chemical facilities are guarded by security guards who are not informed about hazardous chemicals and the risks posed by their theft and diversion. That is because there is a widespread belief that diversion of chemicals for malevolent purposes is highly unlikely. Security concerns do not rise beyond prevention of ordinary theft.

University staff members who are responsible for security of chemical storage are often low-ranking employees who earn very low wages and have little education. In some cases, security personnel have obtained restricted chemicals and used them for illegal purposes. For example, theft of ethanol from chemical stores is common throughout the world.

### Reporting Systems

Lack of good security protocols may result in a failure to report lapses in security. Cultural attitudes may result in an acceptance of the failure to report problems. What constitutes a security breach or problem is often not well defined nor widely disseminated. Persons at all levels of the laboratory do not understand what a reportable incident is and how and to whom to report it.

## CURRENT EDUCATIONAL OUTREACH EFFORTS IN DEVELOPING COUNTRIES

International, regional, and national organizations often act as conduits for supplying chemical information, training, and professional guidance to laboratories in developing countries. Such organizations as the International Union of Pure and Applied Chemistry (IUPAC) and the Organization for the Prohibition of Chemical Weapons (OPCW) have been engaged in those kinds of outreach efforts for many years. The U.S. State Department Chemical Security Engagement Program (CSP) is one of the most active and well-funded programs in place today; it conducts its activities in conjunction with many partnering organizations (including IUPAC and the OPCW) throughout the world.

Because CSP is still in its infancy and is seeking to increase its impact, the Department of State called on the National Academies to assist in the CSP's efforts to promote chemical safety and security in developing countries.[9] In this report, the committee examines the dual-use risks posed by toxic industrial chemicals and other hazardous chemicals and provides guidance on a baseline of practices in the handling and storage of hazardous chemicals required to promote safety and security in their use on a laboratory scale in developing countries. In its second task, the committee will be producing educational materials for CSP training.

### The U.S. Chemical Security Engagement Program

In 2007 the Office of Cooperative Threat Reduction in the U.S. Department of State's Bureau of International Security and Nonproliferation initiated the CSP at Sandia National Laboratories as a component of the Global Threat Reduction programs. These programs are "aimed at reducing the threat posed by terrorist organizations or states of concern seeking to acquire weapons of mass destruction expertise, materials and equipment."[10]

As explained by the State Department official at the first committee meeting, at the forefront of the CSP are the goals of engaging chemical professionals in academia and industry, raising awareness about the threat of chemical dual-use, and fostering national and regional improvement in chemical safety and security. The program also wishes to identify chemical safety and security gaps, promote chemical safety and security best practices, and establish cadres of chemical safety and security officers through training workshops and other outreach efforts.[11]

The CSP engages with countries that are active producers or exporters of industrial chemicals or have growing chemistry capabilities and indus-

---

[9]See Appendix A for the full statement of task.

[10]See *http://www.csp-state.net/* (accessed October 29, 2009).

[11]The educational materials being produced by the committee for CSP to use in its training activities will be delivered separately. They will be based on material generated by this current study (mainly Chapters 3 and 4 and Appendixes F and G) and derived from the forthcoming revised edition of *Prudent Practices in the Laboratory: Handling and Management of Chemical Hazards*. The educational materials will be distributed in print and electronic format, and will be translated into Arabic, French, and Indonesian.

trial and regional security concerns. CSP is currently working with the following countries: Afghanistan, Indonesia, Malaysia, Pakistan, and the Philippines in South and Southeast Asia and Bahrain, Egypt, Iraq, Jordan, Morocco, United Arab Emirates, and Yemen in the Middle East and North Africa. The CSP partners with organizations in those countries and others at the international level (shown in Box S-1).

In academia the CSP seeks to develop and implement training activities to reinforce best practices in chemical security and safety in chemistry curricula. As discussed earlier, universities have unique risks that can include lack of safe practices, presence of COCs, improper management and storage of chemicals, and lack of enforcement of safety rules. Through its training activities, the CSP hopes to prepare laboratories in developing countries to avoid the consequences of chemical mismanagement, such as bodily

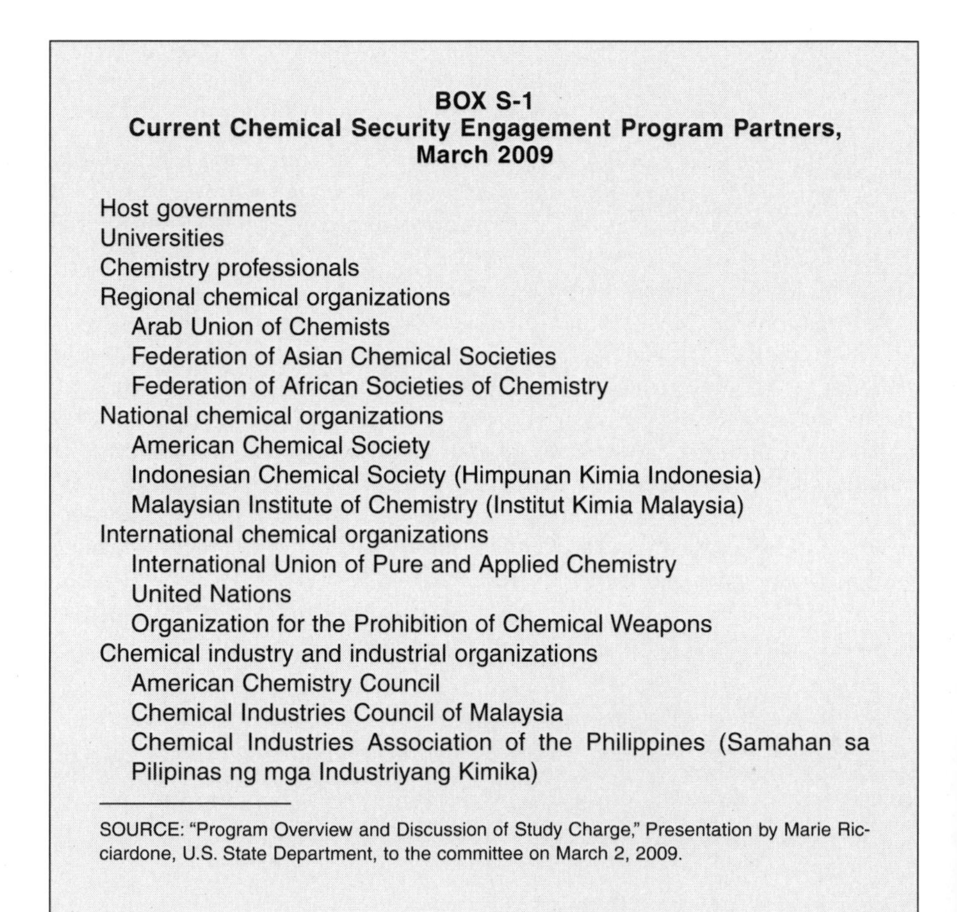

**BOX S-1**
**Current Chemical Security Engagement Program Partners, March 2009**

- Host governments
- Universities
- Chemistry professionals
- Regional chemical organizations
  - Arab Union of Chemists
  - Federation of Asian Chemical Societies
  - Federation of African Societies of Chemistry
- National chemical organizations
  - American Chemical Society
  - Indonesian Chemical Society (Himpunan Kimia Indonesia)
  - Malaysian Institute of Chemistry (Institut Kimia Malaysia)
- International chemical organizations
  - International Union of Pure and Applied Chemistry
  - United Nations
  - Organization for the Prohibition of Chemical Weapons
- Chemical industry and industrial organizations
  - American Chemistry Council
  - Chemical Industries Council of Malaysia
  - Chemical Industries Association of the Philippines (Samahan sa Pilipinas ng mga Industriyang Kimika)

SOURCE: "Program Overview and Discussion of Study Charge," Presentation by Marie Ricciardone, U.S. State Department, to the committee on March 2, 2009.

injuries or expensive clean-up of laboratory waste. The CSP is currently in the process of establishing chemical safety and security officer networks that will develop modules for hands-on training, work with universities to identify candidates for the networks, and offer a five-day train-the-trainer course that focuses on fundamentals of chemical management and elements of chemical safety and security.

The CSP is also partnering with chemical industrial organizations to promote established best practices in chemical security, such as those reflected in the Responsible Care® Security Code and Responsible Care® Management System.[12] Chemical industry risks can include theft of unsecured chemicals and improper disposal of chemicals, in addition to many of the risks that academia faces.

## KEY FINDINGS AND RECOMMENDATIONS

Developing countries have unique needs, and their safety and security practices and attitudes vary considerably, both among developing countries and between developing countries and developed countries. Many in the international community are addressing safety and security in developing countries, but much more attention is needed to better understand safety and security practices and malpractices, and to strengthen and expand outreach efforts. The CSP has undertaken a unique and valuable outreach effort. There is clearly more to do and more opportunities than the program at its current size can address.

The committee recommends that the CSP consider the three main actions outlined below. Chapters 2-4 each provide detailed guidance and are summarized by Recommendations 1-3, respectively. People responsible for implementing safety and security programs will need to read these chapters in their entirety.

1. **Recommendation 1: Build strong relationships** (Chapter 2). The CSP should continue to develop additional and stronger organizational relationships and networks of chemical safety professionals. This includes expanding existing relationships with the groups listed in Box S-1 and creating new ones with international groups such as the International Program on Chemical Safety, U.N. organizations, and other regional and national organizations. Additional considerations for enhancing safety and security include forging private-public partnerships, building on existing certification and training programs, and encouraging and promoting international standards in chemical security and safety.

[12]For more information, see *http://www.americanchemistry.com/s_responsiblecare/sec.asp?CID=1298&DID=4841*.

2. **Recommendation 2: Establish management systems** (Chapter 3). The CSP should provide guidelines for establishing safety and security programs over the life cycle of a chemical, from planning and procurement to ultimate use and final disposal. This should include systematically integrating safety and security into a research institution to anticipate and prevent circumstances that might result in injury, illness, or adverse environmental effects. A critical aspect of such guidelines being successfully implemented is the commitment and support it should have from top leaders in the institution.

3. **Recommendation 3: Comply with rules, programs, and policies** (Chapter 4). The CSP should encourage institutions, organizations, and industries in developing countries to develop clearly defined policies for enforcing and complying with safety and security rules. The policies should include establishing programs for regular inspections, a method for reporting safety and security incidents, investigations, follow-up, enforcement, and systems for reward and recognition, which will require hiring and maintaining the appropriate level of trained safety and security staff (as outlined in Chapter 4). In addition, the CSP should assist laboratory managers in developing countries with identifying noncompliant behaviors and finding ways to address them directly through education and training.

# 1

# Current Patterns of Procurement, Use, and Distribution of Chemicals in Developing Countries

After hearing from guest speakers from developing countries and reviewing publicly available information on global procurement and distribution of laboratory chemicals, the committee was not surprised to find wide variation in developing countries. Many laboratory reagents take circuitous routes to their final destinations. For example, a laboratory chemical may be purchased from a U.S. chemical company, but the network of distribution may include the movement of the chemical from a manufacturing country (such as China) to the United States, then to a distributor in Europe, and finally to a developing country such as the Philippines. At the same time, some chemicals are also directly imported by developing countries from places such as Europe, China, and Japan. In some cases, distributors in a country may keep a stock of chemicals; in others, they obtain chemicals on request. Importation and customs issues can delay delivery of chemicals. It is not unusual for procurement of a single chemical to take as long as six months. There are also concerns about quality; some distributors repackage chemicals or obtain them from questionable sources. Transportation may raise problems; piracy at sea and theft from ground transport constitute risks.

Chemists and other scientists collectively use thousands of chemicals in their laboratory work, but some chemicals pose a particular risk to the general public if they are acquired by people who wish to inflict harm. Such chemicals are commonly known as dual-use or multiple-use chemicals. In this report, the committee has chosen to use the term chemicals of concern (COCs), which includes chemicals listed by the Chemical Weapons Convention, chemicals that have potential for mass destruction, explosives and precursors of improvised explosive devices, and chemicals of high acute toxicity (rated as Category 1 in the Globally Harmonized System of Classification and Labeling of Chemicals). Examples of COCs are provided in Appendix D.

A committee review of the chemical supply chain in developing countries has uncovered the disturbing fact that essentially any chemical can be obtained from suppliers by anyone without much effort. The manufacture, importation, transport, storage, and sale of chemicals are poorly regulated. Even the end-user certificates required for the purchase of some hazardous chemicals can easily be obtained. Thus, academic laboratories and university chemical storage are not the only places from which dual-use chemicals can be obtained.

Chemical laboratory activities in universities, government agencies, and private industry in developing countries are part of a global chemical enterprise. The large-scale manufacture and distribution of commodity and specialty chemicals for commercial purposes dominates the enterprise. However, small-scale chemical laboratory research and teaching activities present vulnerabilities for safety and security. They tend to operate independently, have less government and regulatory oversight, and are more accessible to the public. Characteristics of the global chemistry business are presented in this chapter to provide context for understanding chemical laboratory safety and security practices in developing countries. Safety and security practices are intended to help laboratories carry out their primary functions efficiently, safely, and securely, but improving safety and security is often seen as inhibitory rather than enabling.

## THE GLOBAL BUSINESS OF CHEMISTRY

"Diverse" is the word that best describes the global chemical industry, whose products are used to satisfy daily consumer needs and for applications as wide-ranging as crop protection, disease prevention, and energy production. The chemical industry converts raw materials such as oil, coal, gas, air, water, and minerals into a vast array of substances for use by chemical companies, other industries, and consumers. The wide variety of products range from commodity industrial chemicals used in making other substances to specialty chemicals tailored for unique applications.[1]

The chemical industry is reported to be one of the most regulated of all industries. Nonetheless, there is a persistent lack of hazard information about most chemical substances on the market and the products in which

[1]OECD. *Environmental Outlook for the Chemicals Industry 2001. http:/www.oecd.org/ehs* (accessed May 8, 2010). Paris: Organisation for Economic Co-operation and Development, 2001.

they are used. In the future this will change as a result of the European Union REACH legislation,[2] which will require hazard information on some of the 30,000 chemicals that are available for sale in Europe. There is no accurate account of the total number of chemicals on the market, but one major global supplier of laboratory chemicals and equipment, Sigma Aldrich, reported distributing about 130,000 chemical products (100,000 chemicals and 30,000 equipment products) to approximately 160 countries worldwide in 2008.[3]

The responsibility to provide this hazard information falls largely on the chemical industry. However, although some of the largest industrial firms in the world are chemical companies, a substantial number of chemicals are produced by small- and medium-sized enterprises (SMEs). For example, companies with fewer than 50 employees make 95 percent of the 50,000 chemicals produced in the United States. Governments have only limited interactions with SMEs, and these companies are often not very involved in the discussions on chemical safety.[4] That makes it difficult to assess information related to chemical management and to implement regulatory controls and measures in SMEs.

The difficulty in obtaining public data on the volume, distribution, and use of chemicals by specific countries involved in the global chemical enterprise is reflected in this chapter, in which most of the numbers are derived from secondary sources and are largely aggregated according to region. There is also little publicly available information on the volume and distribution of chemicals used by academic and research laboratories, especially those in developing countries. The analysis of supply and distribution to academic laboratories in particular is therefore supplemented by a bibliometric analysis of current chemical literature.

---

[2]Registration, Evaluation, Authorization and Restriction of Chemicals (REACH) is a new European Community regulation on chemicals and their safe use (EC 1907/2006) that entered into force on June 1, 2007. See *http://ec.europa.eu/environment/chemicals/reach/reach_intro.htm* (accessed January 12, 2010).

[3]Creating Differentiation Through Innovation. Sigma Aldrich Annual Report. 2008. *http://www.sigmaaldrich.com/site-level/corporate/annual-report-2009/arhome-2008.html* (accessed January 22, 2010) and 2008 Form 10-K filing to the Securities and Exchange Commission: *http://www.sec.gov/Archives/edgar/data/90185/000119312509041007/d10k.htm* (accessed January 22, 2010).

[4]OECD. *Environmental Outlook for the Chemicals Industry 2001. http:/www.oecd.org/ehs* (accessed May 8, 2010). Paris: Organisation for Economic Co-operation and Development, 2001.

## CHEMICAL SUPPLY

In 2008 the global chemical industry was a $3.7 trillion enterprise.[5] In the past, the United States and Western Europe were the top exporters of chemicals to developing countries, but now they lag behind Asia-Pacific countries, primarily because of production in China and India. Surpassing U.S. and European output, the Asia region experienced an increase of 9.1 percent in its share of world chemical sales (Figure 1-1) from 1997 (17.0 percent) to 2007 (30.4 percent), a stark contrast with the decline in both Europe (from 32.2 percent to 29.5 percent) and the countries adhering to the North American Free Trade Agreement (from 28.0 percent to 22.2 percent).[6]

Investment in new plants and equipment reflect a preference for the Asia-Pacific region. The American Chemistry Council (ACC) reports a growth in global capital investment from 2006 ($171 billion) through 2009 ($237 billion), and the Asia-Pacific region (excluding Japan) accounted for 56 percent of the gain in the period.[7] In comparison, the United States accounted for only 6 percent of the gain. The shift to the Asia-Pacific region is attributed largely to China's increasing share of global chemicals production.[8]

According to the World Bank, developing countries are characterized as having low- to lower-middle-income economic status,[9] but they can vary widely in socioeconomic standards and implementation of the rule of law, and they are neither major consumers nor producers of chemicals in global terms. Petrochemical commodities—polymers and fertilizers—are the main products of the developing countries' industries. However, it is predicted that a shift in chemical production from developed countries making up the Organisation for Economic Co-operation and Development (OECD) to non-OECD countries will take place within the next 10 years. A number of developing countries have the capability for increasing pharmaceuticals production and many are investing in oil and gas, which are key drivers for

[5]American Chemical Council. 2009. *Guide to the Business of Chemistry*. *http://www.americanchemistry.com* (accessed October 23, 2009).

[6]CEFIC (European Chemical Industry Council). 2009. *Facts and Figures*, Chapter 1: Profile of the Chemical Industry. *http://www.cefic.org/factsandfigures/* (accessed February 11, 2010).

[7]American Chemical Council. 2009. *Guide to the Business of Chemistry*. *http://www.americanchemistry.com* (accessed October 23, 2009).

[8]CEFIC (European Chemical Industry Council). 2009. *Facts and Figures*, Chapter 1: Profile of the Chemical Industry. *http://www.cefic.org/factsandfigures/* (accessed February 11, 2010).

[9]See the World Bank country classifications: *http://data.worldbank.org/about/country-classifications* (accessed May 6, 2010).

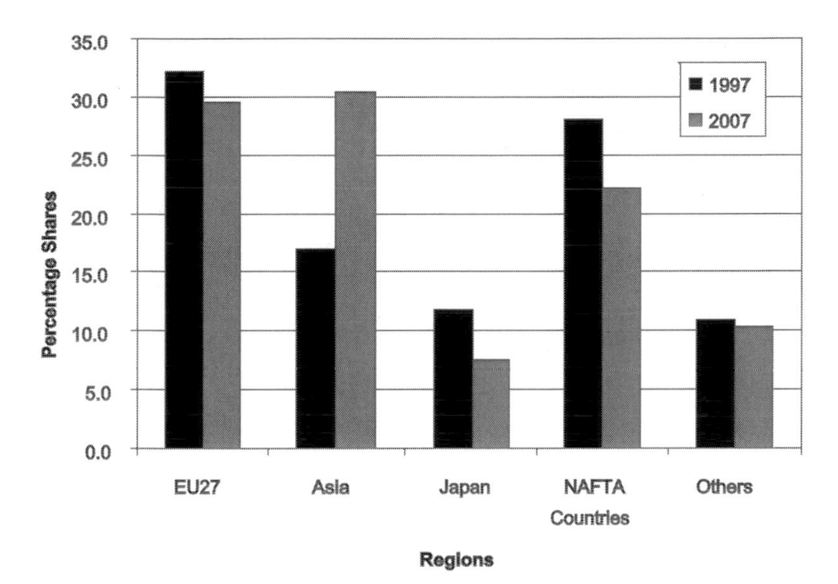


FIGURE 1-1 Sales of chemicals by region where sold: 1997 vs. 2007.
NOTE: "Asia" does not include Japan.
SOURCE: Cefic.

the chemicals industries. Important developing-country producers include India, China, Indonesia, and the Philippines.[10]

The production of agricultural chemicals is a focus of the chemical industry in developing countries. These countries accounted for 5 percent of total world nitrogenous fertilizer production in 2002. Developing countries also contributed about 4 percent to total world production of pesticides (insecticides, fungicides, and disinfectants) in 1998, and about 5 percent in 2002.[11] Although the current focus of the chemical market in Africa is on meeting local needs, the proximity, particularly for North African nations, to European markets has led to a greater focus on exporting. Morocco, for

---

[10]OECD. *Environmental Outlook for the Chemicals Industry 2001. http:/www.oecd.org/ehs* (accessed May 8, 2010). Paris: Organisation for Economic Co-operation and Development, 2001.

[11]N. Manda and J. Mohamed-Katarere. 2006. Chemicals. In *Africa Environmental Outlook-2: Our Environment, Our Wealth*. Nairobi: United Nations Environmental Program. *http://www.unep.org/DEWA/Africa/docs/en/AEO2_Our_Environ_Our_Wealth.pdf* (accessed October 23, 2009).

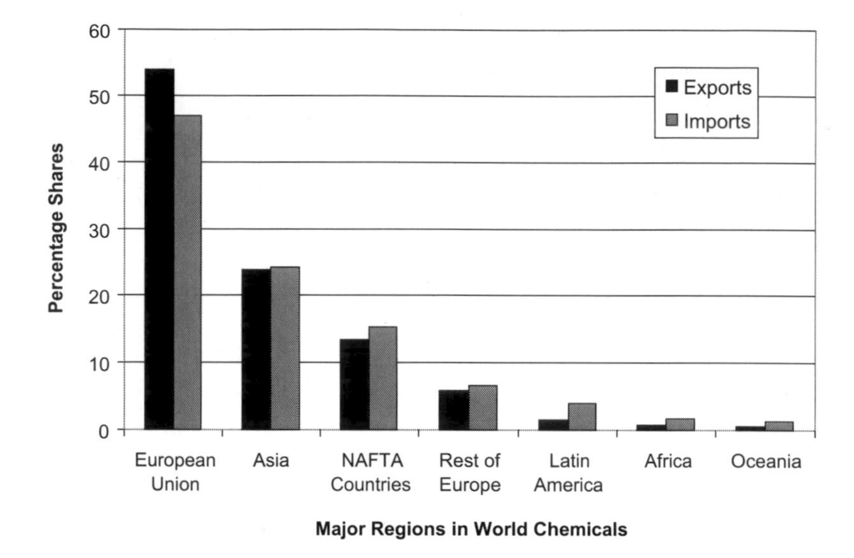


FIGURE 1-2 Regional shares of world exports and imports of chemicals, for 2007.
SOURCE: Cefic.

example, has two-thirds of the world's reserve of phosphate rock and is a leading producer of phosphoric acid.[12]

## CHEMICAL DISTRIBUTION

### Geographic Distribution

Patterns of distribution and customer identification are difficult to obtain from public information. The information used in this chapter comes primarily from the Cefic: European Chemical Industry Council and ACC reports. As mentioned earlier, in terms of regional sales of chemicals (Figure 1-1), Asia (other than Japan) accounts for the largest share: 30.4 percent of €1.82 trillion total world chemical sales in 2007.

Export and import figures (Figure 1-2) for 2007 still point to the European Union as the world leader, accounting for half the global trade.

The top 100 chemical distributors have their main stocking locations in the United States and Canada with an average of 195 employees, and

[12]K. Temsamani. 2009. Presentation at Meeting 3, Committee on Promoting Safe and Secure Chemical Management in Developing Countries, The National Academies, Washington D.C. (see Appendix C).

they import an average of 21 percent of the chemicals that they sell.[13] In 2005 they had on average 11.8 stocking locations around the world, including Europe (15 percent), Mexico (11 percent), Asia (9 percent), and Latin America (7 percent). Those distributors had an average of 3.7 million gallons of bulk storage capacity, an average of 377,000 square feet of warehouse space, and used various transport methods including trucks, vans, trailers, tankers, and railcars.

As distinct from distributors, suppliers also have a global presence. Sigma-Aldrich for example has over 7,900 employees in 37 countries and reported distribution of its total sales as 43 percent in Europe, 35 percent in the United States, and 20 percent in Canada, Asia-Pacific, and Latin America combined.[14] There are also confirmed reports that Sigma-Aldrich chemicals have reached Morocco, possibly through resellers.

The Responsible Care® Status Report 2008 published by the International Council of Chemical Associations lists 2,421 distribution "incidents" (accidental chemical releases during transport of goods by various means) in 26 (out of 53) countries in 2006. The report showed a general decline in the number of incidents since 2001, when there were 3,827 incidents reported. In the same period, there was a 15 percent increase in materials distributed.[15]

## Sector Distribution

Most chemical sales are to such industries as paints and coatings, cosmetics and personal care, foods and beverages, adhesives and sealants, and soaps and detergents.[16] Specific data that would clarify the overall distribution of chemicals to academic and research laboratories are not available. Inferences can be derived from information gleaned from corporate reports of chemical suppliers. For example, in 2008, academic laboratories, government agencies, and nongovernment organizations accounted for 26 percent of Sigma-Aldrich's customers; pharmaceutical companies (35 per-

[13]S. Avery. 2006. The Top 100 Chemical Distributors: Demand remains strong despite high prices. *http://www.purchasing.com/article/219264-The_Top_100_Chemical_Distributors_Demand_remains_strong_despite_high_prices.php* (accessed October 30, 2009).

[14]Creating Differentiation Through Innovation. Sigma Aldrich Annual Report. 2008. *http://www.sigmaaldrich.com/site-level/corporate/annual-report-2009/arhome-2008.html* (accessed January 22, 2010) and 2008 Form 10-K filing to the Securities and Exchange Commission *http://www.sec.gov/Archives/edgar/data/90185/000119312509041007/d10k.htm* (accessed January 22, 2010).

[15]International Council of Chemical Associations. 2009. Responsible Care Status Report 2008. *http://www.responsiblecare.org/filebank/Status%20Report%2001_05.pdf* (accessed January 26, 2010).

[16]American Chemical Council. 2009. *Guide to the Business of Chemistry. http://www.americanchemistry.com* (accessed October 23, 2009).

cent), chemical industries (31 percent), and hospitals and commercial laboratories (8 percent) complete the list. Sigma-Aldrich's reported $2.2 billion in sales and a net income of $342 million in 2008 but did not report the corresponding quantities of chemicals sold.

### Customer Screening

Increased sales in non-European markets have been bolstered by online transactions. For example, Web-based sales accounted for 42 percent of all Sigma-Aldrich Research sales in 2008.[17] The trend is illustrated by the emergence of e-commerce portals of major suppliers and distributors. Products can be ordered online, but although all requests are supposed to be reviewed to verify membership in an organization, registering an online profile does not require such information as the organization's name and address. The process does typically require proof of an established relationship with a local distribution branch through an existing account number.

Such processes are part of the chemical industry's steps to improve chemical safety and security. Responsible Care® is the chemical industry's global, voluntary initiative aimed at improving the health, safety, and environmental effects of the industry's products and processes. According to an interaction during a recent seminar on chemical safety and security, it was stated that although large industries have established protocols (Figure 1-3), local distributors and middlemen have the ultimate responsibility for customer screening.[18,19] At the same time, middlemen may skirt laws, have control over the market prices, hoard or adulterate chemicals, and carry out other malpractices.

The problem, especially in developing countries, is that distributors cannot ensure that their customers will not resell the purchased chemicals. Chemicals are widely traded, even in hardware stores and grocery stores, and regulations for the control of biological, radioactive, and nuclear materials are often inappropriate for chemicals.[20] For example, a local reseller in

---

[17]Creating Differentiation Through Innovation. Sigma Aldrich Annual Report. 2008. *http://www.sigmaaldrich.com/site-level/corporate/annual-report-2009/arhome-2008.html.* (accessed January 22, 2010) and 2008 Form 10-K filing to the Securities and Exchange Commission *http://www.sec.gov/Archives/edgar/data/90185/000119312509041007/d10k.htm* (accessed January 22, 2010).

[18]T. Cromwell. 2009. Leveraging the Relationship between Chemical Safety and Chemical Security to Reduce Terrorism Threats. Presentation given at Asia Pacific Seminar on Chemical Safety and Security to Counter Terrorism, Canberra, Australia.

[19]R. Batungbacal. 2009. Asia Pacific Seminar on Chemical Safety and Security to Counter Terrorism, Canberra, Australia. Personal communication.

[20]R. Mathews. 2009. Lessons Learned from Nuclear, Biological, and Radiological Experiences. Presentation given at Asia Pacific Seminar on Chemical Safety and Security to Counter Terrorism, Canberra, Australia.

FIGURE 1-3 Information/cyber security.
SOURCE: Courtesy of The Dow Chemical Company © 2006.

the Philippines replaces chemical labels, using generic names to avoid regulations on end-user certificates and Materials Safety Data Sheets (MSDSs), and sells the products to the general public. In such a case, the identity and responsibility of the eventual customer can no longer be ascertained by the distributor, let alone by a supplier who is practicing Responsible Care®.

In the case of the release of the nerve agent sarin in the subway of Tokyo, Japan, on March 20, 1995, the Aum Shinrikyo cult responsible for the attack set up a dummy company to purchase the chemical precursors used in the production of the toxin. The facility where the sarin was synthesized escaped zoning and environmental authorities and was discovered only during the investigation of the incident.[21]

Vigilant suppliers can thwart malicious intentions from the very beginning. For example, an alert employee of a distributor in the Northern Territory of Australia, informed law-enforcement authorities when clients

[21]N. Aoki. 2009. Sarin Gas Attacks. Presentation given at Asia Pacific Seminar on Chemical Safety and Security to Counter Terrorism, Canberra, Australia. See also *www.cdc.gov/ncidod/eid/vol5no4/olson.htm* (accessed October 23, 2009).

in Sydney ordered nitric acid, sulfuric acid, glycerin, and other chemicals.[22] The clients intended to synthesize 40 kg of nitroglycerin to blow up and steal from ATMs and bank safes.[23] Distributors, in general, are watchful for suspicious customer behavior, including excessive payments for rapid delivery, use of cash payments instead of charges to a corporate account, delivery to post office boxes, orders for unusual quantities, and insistence on unmarked packaging.[24]

## LABORATORY-SCALE USE OF CHEMICALS

In the absence of publicly available data on the distribution of chemicals (especially COCs) to academic and research laboratories, the committee analyzed articles in scholarly journals to gain some insight into the patterns of use and distribution of laboratory chemicals in developing countries. The search (using the Web of Science database) was conducted for four countries where the Chemical Security Engagement Program is engaged: Indonesia, Malaysia, Pakistan, and the Philippines. It was limited to articles published in 2008 and articles whose correspondence addresses were only in those countries. An initial refinement specified fields of chemistry, for example, multidisciplinary, medicinal, physical, inorganic. A list of chemicals was compiled by using the keywords and abstracts of the articles.

Pakistan is by far the most prolific publisher of the four countries with 412 articles compared with 166 from Malaysia, 10 from Indonesia, and 5 from the Philippines. Of the papers from Pakistan and Malaysia, 44 (11 percent) and 11 (6.5 percent), respectively, dealt with COCs. One of the articles from Pakistan used trinitrotoluene, and many articles involved heavy metals and some pesticides, although most of the chemicals written about were common organic solvents such as acetone, listed in Table D-5, "Chemicals Used in Clandestine Production of Illicit Drugs," in Appendix D.

---

[22]V. Otieno-Aligo. 2009. Application of Forensic Chemistry to Criminal Offences Involving the Use of Chemical Agents. Presentation given at Asia Pacific Seminar on Chemical Safety and Security to Counter Terrorism, Canberra, Australia.

[23]G. Jones. 2007. Police Arrest Bomb Plotters Ahead of APEC. *http://www.dailytelegraph.com.au/news/nsw-act/sydney-bomb-plotters-seized/story-e6freuzi-1111114284160.*

[24]M. Donnan. 2009. Strengthening Chemical Security: An industry perspective. Presentation given at Asia Pacific Seminar on Chemical Safety and Security to Counter Terrorism, Canberra, Australia.

# 2

# Chemical Laboratory Safety and Security Outreach in Developing Countries

Many national and international chemical societies have a strong interest in chemical safety and security, and some already have internal committees for safety in place. Those societies in developing countries share many concerns regarding safety and security in laboratories with more developed countries. They all attempt to develop materials to promote safety but often in isolation from other chemical societies. The U.S. Chemical Security Engagement Program (CSP) plays an important role in fostering interactions with and between chemical organizations that focus on safety and security issues.

The first CSP outreach was a security and safety workshop held in 2007 at the 12th Asian Chemical Congress in Kuala Lumpur, Malaysia, in partnership with the International Union of Pure and Applied Chemistry (IUPAC), the Federation of Asian Chemical Societies (FACS), and the Malaysian Institute of Chemistry (Institut Kimia Malaysia). Since then, the CSP has partnered with many other organizations at the international, regional, national, and local level to conduct similar workshops and training sessions in additional countries, including the Philippines, Thailand, Indonesia, Jordan, and Egypt. There are, however, ways the CSP might expand and build on the work of current partner organizations, and build new relationships to better leverage its current outreach. This chapter discusses some of the current CSP partners and outreach efforts, recommends opportunities for expanding those relationships and fostering new ones, and some potential new approaches to outreach in developing countries.

## EXPANDING CURRENT CSP PARTNERSHIPS

### International Union of Pure and Applied Chemistry

IUPAC is a global organization of more than 50 national adhering organizations (NAOs), and additional associate NAOs, associated organizations, company associates, and affiliated members.[1] Typically the different adhering organizations are national scientific or chemical societies, such as the Chemical Society of Ethiopia and the Chemical Society of Pakistan. In the United States, however, the NAO is the U.S. National Academy of Sciences. The strategic goals of IUPAC include providing leadership in addressing global issues in the chemical sciences, emphasizing the needs of chemists in developing countries, and using its global network to contribute to the advancement of chemistry education, which aligns well with CSP efforts to engage chemical professionals on laboratory safety and security.

#### Safety Training Program

The IUPAC project most directly related to outreach with respect to chemical laboratory safety and security is the Safety Training Program.[2] This unique program enables experts from developing countries to learn about safety and environmental protective measures by visiting and working in chemical plants of IUPAC company associates in the industrialized world. Safety Training Program trainees are (1) professional scientists and engineers who are working at supervisory or managerial levels in chemical companies, government or scientific institutions, or universities; (2) engaged in aspects of safety, security, or environmental protection or in the teaching of these fields; and (3) are influential in their workplaces and their home countries.

Trainees spend two to four weeks in intensive "shadowing" of health, safety, and environmental professionals in their host companies, learning state-of-the-art techniques and practices and participating in meetings and decision-making functions. They write reports on their training and plans for implementation of new initiatives at home, and have the opportunity to participate in regular Safety Training Program workshops at IUPAC congresses. In the workshops, trainees share best practices, present their work, and hear invited speakers on topics of concern in health, safety, and the environment. In the future, the CSP could collaborate with the Safety Training Program to make these workshops more than presentations and

[1] For more information, see *http://www.iupac.org* (accessed October 26, 2009).

[2] For more information, see *http://www.iupac.org/standing/coci/safety-program.html* (accessed October 26, 2009).

provide training in exhibiting and demonstrating equipment for safety and security.

### Educational Resources on the Web

IUPAC has a variety of resources on its Web site that are relevant to laboratory safety and security, including teaching modules and other resources on toxicology,[3] environmental monitoring, and its collaboration with the U.N. Educational, Scientific, and Cultural Organization (UNESCO) on the Global Microscience Project.[4]

The IUPAC project Multiple Uses of Chemicals, in conjunction with the Organization for the Prohibition of Chemical Weapons (OPCW), has developed educational material for chemists and chemistry teachers about the Chemical Weapons Convention (CWC).[5] The materials are available on the Internet,[6] and discuss beneficial uses of chemicals, and possible misuses of chemicals, including the production of chemical weapons.

### The International Year of Chemistry in 2011

The International Year of Chemistry (IYC) in 2011 came about through a partnership between IUPAC and UNESCO.[7] Regional, national, and local chemical societies; organizations; teaching and research institutions; governments; and industry will collaborate in a series of events to celebrate the accomplishments and future of chemistry during IYC 2011.

While issues of safety and security are not included in the specific goals of IYC 2011 celebrations, the increased attention that will be given to chemistry topics during IYC 2011 provides an excellent opportunity for outreach on chemical laboratory safety and security, including the misuses of chemicals.

### IUPAC Partners

Members of IUPAC are representatives of many other organizations, some of which could facilitate the development of an organizational network focused on safety and security. The fellows of the IUPAC Safety Training Program can be approached to assist in dissemination and use of

---

[3]For more information, see *http://www.iupac.org/divisions/VII/VII.C.2/index.html* (accessed October 26, 2009).

[4]For more information, see *http://www.iupac.org/publications/ci/2002/2403/smallscale chemistry.html* (accessed October 26, 2009).

[5]Alastair Hay, co-chair of the project, presented to the committee on June 4, 2009.

[6]For more information, see *http://multiple.kcvs.ca/* (accessed October 26, 2009).

[7]For more information, see *http://www.chemistry2011.org* (accessed October 26, 2009).

educational materials. The fellows were chosen for their ability to interact effectively in their places of employment, their scientific societies, and local, regional and national governments to improve health, safety, and environmental practices in their home countries.

### Organization for the Prohibition of Chemical Weapons

The OPCW[8] is the implementing body for the CWC, which underlies the outreach efforts of the CSP. There are 188 nations (98 percent of the global population) that have joined the OPCW. The OPCW maintains a Web site with a list of scheduled chemicals,[9] an important resource for instructors and other persons responsible for maintaining laboratory safety and security. OPCW provides funding for its officers to speak at important conferences to introduce the OPCW and CWC to encourage involvement by the chemical community. The OPCW also provides grants for instrumentation and research to foster the peaceful applications of chemistry.

OPCW has collaborated with IUPAC to hold two workshops on the CWC. The workshops have helped to supply OPCW with information on new and emerging science and technology related to the synthesis of multiple-use chemical materials and to the detection, analysis, and destruction of chemical weapons. In addition, as mentioned earlier, OPCW has partnered with IUPAC on chemical education and outreach aimed at informing students about chemical weapons and the CWC.[10]

### American Chemical Society

The American Chemical Society (ACS) is the world's largest scientific society, with more than 154,000 members, 19,000 of whom are in over 100 countries outside the United States. ACS sponsors or promotes many international activities, such as joint conferences with chemical societies in other countries. For example, the International Chemical Congress of Pacific Basin Societies (PacifiChem) is a weeklong scientific meeting, held every five years in conjunction with ACS counterparts in Australia, Canada, Japan, Korea, New Zealand, and China. ACS activities are driven largely by members through its committees, two of which are described below.

---

[8]For more information, see *http://www.opcw.org/* (accessed October 26, 2009).

[9]For more information, see *https://apps.opcw.org/cas/* (accessed October 26, 2009).

[10]For more information, see *http://www.iupac.org/web/ins/2005-028-1-050* (accessed October 26, 2009).

### ACS Committee on International Activities

The ACS Committee on International Activities (IAC), supported by the staff in the Office of International Activities, helps scientists and engineers worldwide to communicate and collaborate. The IAC is active in conferences, symposia, and workshops around the globe; other work includes facilitating visas for travel to the United States, providing international news to the scientific community, and operating a free online network for scientists in a number of languages. Laboratory safety and security are not topics of strong interest for this committee, but this could change with increased expressions of interest from ACS members.

### ACS Committee on Chemical Safety

The prime responsibility of the ACS Committee on Chemical Safety (CCS) is the encouragement of safe practices in chemical activities. The CCS serves as a resource for chemical professionals mainly in the United States in providing advice and counsel on the handling of chemicals, and it seeks to ensure safe facilities, designs, and operations by calling attention to potential hazards and stimulating education in safe chemical practices. The CCS also provides advice to other ACS units on matters related to chemical safety and health. A variety of publications are available free on its Web site.[11]

### ACS Division of Chemical Health and Safety

The ACS Division of Chemical Health and Safety (CHAS) is an international organization with about 1,300 members dedicated to advancing health and safety in the chemical enterprise.[12] CHAS provides technical programming at ACS national and regional meetings, produces its own journal (*Journal of Chemical Health and Safety*), and has an active Listserv to benefit members worldwide who have questions about safety issues. It has members in a number of developing countries who occasionally post questions on the Listserv.

Safety materials produced by ACS and its committees and divisions are available in English and sometimes in other languages. For example, the publication *Safety in Academic Chemistry Laboratories* (SACL) is used widely abroad; the Spanish translation is used in South and Central America. The CCS also has several members who have engaged in safety training

---

[11]For more information, see *http://membership.acs.org/C/CCS* (accessed October 26, 2009).

[12]For more information, see *http://membership.acs.org/c/chas/default.htm* (accessed October 26, 2009).

in developing countries; however, their initiatives are not tied directly to the CCS or CHAS. In addition, the CCS has recently appointed a liaison to the IAC; this may provide for future collaboration between the committees on chemical safety practices internationally.

### Other ACS Opportunities

ACS has the potential to play more of a leadership role in developing better communication on safety and security between chemical societies throughout the world. ACS has enormous resources, especially its strong volunteer committees that could be expanded to facilitate worldwide communication between chemical organizations.

The ACS and IUPAC are especially valuable partners for CSP in helping to create networks of chemical safety professionals, inasmuch as each organization has a global reach and effective contacts with national chemical societies and related organizations around the world.

## Regional Organizations

Regional organizations, such as the Arab Union of Chemists, Federation of Asian Chemical Societies, and the Federation of African Societies of Chemistry (FASC), are key CSP partners in outreach to developing countries. For example, the president of FASC, Temechegen Engida, spoke to the committee on his collaboration with the Pan African Chemistry Network to conduct a recent survey of students and instructors at his university and other institutions in Africa. Clear indications for the need for improvements in instruction, equipment, procedures, and infrastructure were found. As a result, FASC plans a series of major events across Africa in 2011, in conjunction with IYC 2011, to promote safe practices and applications of chemistry in laboratories, chemical production, and end uses. Major resource needs were also identified, including funding, communication infrastructure, trained personnel, and institutions willing to commit to the regional efforts.

## The American Chemistry Council

The American Chemistry Council (ACC) is the trade association for the U.S. chemical industry and has global influence because of the international nature of the chemical industry. ACC is well known for its Responsible Care® program,[13] which is a global initiative established in 1988 focused

[13]For more information, see *http://www.americanchemistry.com/s_responsiblecare/sec.asp?CID=1298&DID=4841* (accessed October 27, 2009).

on advancing safe and secure management of chemical products and processes. Currently, 53 national chemical industry associations participate in the program. In 2001, the Responsible Care® Security Code (RCSC) was added to the program. The RCSC requires member companies to conduct comprehensive security vulnerability assessments, implement security enhancements, and obtain independent verification that those enhancements have been made. It also requires companies to create security management systems, which are documented to provide quality control and assurances. All the details of the RCSC are posted on the ACC Web site and are publicly available, including detailed guidance for implementing it. The tools on the Web site are excellent resources for any facility, including an academic laboratory that intends to assess security risks and implement a robust facility security program.

Many ACC members are large multinational corporations, and they implement the Responsible Care program globally. Such companies also provide outreach and training for local universities and smaller laboratories. The ACC has worked to support the CSP in numerous parts of the world—including Malaysia, the Philippines, Pakistan, Afghanistan, Indonesia, Vietnam, Thailand, and the United Arab Emirates—over the last three years. It works with local chemical associations to facilitate the development of safety and security practices tailored to local needs. In addition, the ACC led the development of a safety and security workshop in conjunction with the 10th anniversary of OPCW.

## POTENTIAL NEW ORGANIZATIONAL RELATIONSHIPS AND APPROACHES TO OUTREACH IN DEVELOPING COUNTRIES

The CSP is partnering with many organizations, but there is a need to continue to develop new organizational relationships and to create networks of chemical safety professionals. This section offers guidance on new organizations that the CSP could partner with and new approaches that the CSP could use to better leverage its outreach.

### International Organizations

Among notable organizations that are capable of outreach to developing countries with respect to chemical laboratory safety and security are the International Program on Chemical Safety (IPCS)[14] and such U.N. organizations as UNESCO. The IPCS INCHEM program, in collaboration with the Canadian Centre for Occupational Health and Safety, provides

[14]For more information, see *http://www.inchem.org/* or *http://www.who.int/ipcs/en/* (accessed October 26, 2009).

resources such as health and safety guides and international chemical safety information cards.[15] UNESCO[16] has been a sponsor of the IUPAC Safety Training Program, has regional offices throughout the world, and has a long record of support for educational initiatives in the sciences, such as the Global Microscience Project. As mentioned earlier, UNESCO is also a partner with IUPAC in the IYC 2011.

The Strategic Approach to International Chemicals Management (SAICM)[17] is another initiative, and is implemented through the U.N. Environment Programme and the World Health Organization. It is a global policy framework for fostering the sound management of chemicals. The SAICM supports the achievement of the goal agreed on at the 2002 Johannesburg World Summit on Sustainable Development: to ensure that by 2020, chemicals are produced and used in ways that minimize substantial adverse effects on the environment and human health. It particularly targets the developing world. The SAICM is still in its formative stages, but it may be useful to monitor its work to judge whether it can play a greater role in promoting safe and secure laboratory practices.

The U.N. Industrial Development Organization[18] manages a network of cleaner production centers throughout the developing world that have responsibilities beyond chemical production and research; they have influence in their regions with respect to regulations, policy, and training. The centers can be contacted to explore the possibility of collaborating on laboratory safety and security training and education.

The International Organization for Chemical Sciences in Development (IOCD) is a U.S.-based nongovernmental organization that supports the sciences in developing countries, with a focus on Africa.[19] The IOCD believes that increased international collaboration will improve the chemical sciences, and in turn, the health and economic status of developing countries. The organization supports symposia, international research sabbaticals, and workshops in laboratory techniques.

---

[15]For more information, see *http://www.inchem.org/pages/icsc.html* or *http://www.cdc.gov/niosh/ipcs/icstart.html* (accessed October 26, 2009).

[16]For more information, see *http://portal.unesco.org/science/en/ev.php-URL_ID=5572&URL_DO=DO_TOPIC&URL_SECTION=201.html* (accessed October 26, 2009).

[17]For more information, see *www.saicm.org* (accessed October 26, 2009).

[18]For more information, see *www.unido.org* (accessed October 26, 2009).

[19]For more information, see *www.iocd.org/vision_mission.shtml* (accessed October 26, 2009).

## Regional and National Organizations

The CSP can play a much greater role in building capacity in developing countries for chemical laboratory safety and security by conducting training programs in collaboration with local or regional chemical societies. Training in developing countries should be arranged with local partner organizations to establish local ownership of the initiatives. The CSP should initiate some programs that focus on motivating institutional leadership. Establishing such relationships with regional and national organizations would enable CSP to increase its impact through organization of more regional conferences.

Developed countries in Asia have the potential to serve as excellent regional partners in reaching out to nearby developing countries. For example, the Council of Scientific and Industrial Research, largely through India's national laboratories, has conducted training programs for young scientists in the fundamentals of laboratory health and safety and laboratory design. The Indian Ministry of Science and Technology also has a Task Force on Green Chemistry, whose mandate includes workshops and interactive sessions for industrial scientists on chemical safety and security. Neither the Indian Chemical Society nor the Chemical Research Society of India has committees on safety and security.

The Chemical Society of Japan (CSJ) has a committee on environment and safety issues, some of whose activities are related to chemical and laboratory safety. The CSJ holds an annual two-day seminar, "Chemical Safety Schooling," in which 70 to 90 representatives of industry and institutes participate. The seminar covers potential risks posed by handling chemicals, preventive measures, and official regulations. During its annual meetings, the committee also conducts a symposium related to laboratory safety. In these symposia, information is shared on recent changes in regulations, laboratory incidents and their analysis, improvements, and examples of teaching materials. A visit to a large chemical company is made to observe safety and environmental facilities in a modern industrial laboratory.

Pakistan has two professional chemistry organizations, the Chemical Society of Pakistan and the National Core Group in Chemistry (NCGC), which are both generously funded by the government of Pakistan. The NCGC has an extensive training program in chemical laboratory safety that includes the training of chemists and research scholars and the publication of laboratory safety handbooks and pictorial booklets. The Ministry of Environment of Pakistan regularly conducts stakeholder consultations and workshops on the implementation of the SAICM in Pakistan.

Another useful resource is alumni associations of U.S. and European universities. For example, members of the alumni association of the United Kingdom's Imperial College or the Massachusetts Institute of Technology alumni association could be approached to find volunteers who could be an advocate for chemical laboratory safety and security. These groups could hold citywide university seminars, workshops, and exhibits in collaboration with the CSP and experts from developed countries, institutes, and vendors of safety equipment.

### Professional Organizations

Three safety-related professional organizations in the United States could be beneficial partners in outreach to developing countries. They can be approached to provide forums for networking, administering comprehensive education programs, and serving as a bridge between scientists and chemical safety officers. Similar organizations that provide comparable information exist in many other countries, such as the British Occupational Hygiene Society.

- The American Industrial Hygiene Association (AIHA)[20] is an organization of professionals dedicated to the anticipation, recognition, evaluation, and control of safety, health, and environmental factors that arise in the workplace and could result in injury, illness, or impairment or affect the well-being of workers and members of the community. The AIHA International Affairs Committee addresses such international issues as continuing education, membership, publications, laboratory accreditation, and humanitarian outreach. AIHA works to establish associations and graduate programs in industrial hygiene in developing countries, and it sponsors symposia, roundtables, and forums on international occupational hygiene issues. It also establishes informational Web sites and Listservs. AIHA has local chapters outside the United States, including a particularly active one in the Middle East Arabian Gulf region.
- The American Society of Safety Engineers (ASSE)[21] is a professional safety organization that serves safety, health, and environmental professionals around the globe. ASSE provides technical information and global and local networking and cooperates with numerous safety, health, and environmental organizations worldwide. ASSE has members in over 64 countries, including Saudi Arabia, Kuwait, Australia, Nigeria, Papua-New Guinea, the United Kingdom, Ecuador, and Egypt.
- The Campus Safety Health and Environmental Management Association (CSHEMA)[22] "provides information sharing opportunities, continuing education, and professional fellowship to people with environmental

[20]For more information, see *www.aiha.org* (accessed October 26, 2009).
[21]For more information, see *www.asse.org* (accessed October 26, 2009).
[22]For more information, see *www.cshema.org* (accessed October 26, 2009).

health and safety responsibilities in the education and research communities." CSHEMA's vision is to have "excellent health, safety, and environmental protection understood and integrated into teaching, research, and service throughout the educational and research communities." Membership in CSHEMA is on an institutional basis. Membership fees are based on the number of employees working in the institution's health and safety group. Fees are waived for colleges and universities in developing nations. CSHEMA presents "webinars" on emerging issues, holds regional conferences, and has an annual international conference on campus health and safety. The CSHEMA Forum provides for online networking, sharing of training materials, posting questions, and other opportunities for collaboration among individuals interested in health, safety, and environmental protection.

### U.S. Government Agencies

Many agencies in the United States engage in activities that could benefit the CSP. For example, the National Institute for Occupational Safety and Health (NIOSH) provides materials and information of interest to persons working on improving chemical laboratory safety and security practices, such as *School Chemistry Laboratory Safety Guide*.[23] A primary focus of its efforts is dual-use chemicals, and its Web site provides information and resources on chemical safety, evaluation of workplace hazards, and training. Another example is the U.S. Food and Drug Administration (FDA). Through its rules on Good Laboratory Practices and Good Manufacturing Practices,[24] the FDA oversees the safety, traceability of production, quality assurance, and others aspects of U.S. pharmaceutical companies operating inside and outside the United States.

## ADDITIONAL CONSIDERATIONS FOR ENHANCING SAFETY AND SECURITY TRAINING

### Private-Public Partnerships

The entire chemical industry is increasingly affected by global conditions. The United Nations, the European Union (EU), and several countries have enacted laws or developed guidance related to the safe and secure management of chemicals. Their initiatives have a global impact. A no-

---

[23]For more information, see *http://www.cdc.gov/niosh/docs/2007-107/pdfs/2007-107.pdf* (accessed January 22, 2010).

[24]See the FDA website for more information: *http://www.fda.gov/Drugs/GuidanceComplianceRegulatoryInformation/Guidances/ucm064971.htm* (accessed January 26, 2010).

table example is the EU chemical law REACH (Registration, Evaluation, Authorization, and Restriction of Chemical Substances), which requires companies that wish to do business in the EU to submit chemical information and assessments.[25]

In response to REACH and other initiatives with a global impact, the private sector has sought to improve communication and coordination within the industry and with government agencies and academia. The global chemical industry is interconnected through the International Council of Chemical Associations (ICCA), which serves as the "world-wide voice of the chemical industry, representing chemical manufacturers and producers all over the world." ICCA members—over 50 national chemical manufacturing associations—also adhere to the Responsible Care program. A key element of Responsible Care is the establishment of partnerships and alliances, among global and regional institutions and at the local level. An example of a successful private-public partnership mentioned earlier in this chapter is ACC's support of the CSP.

Companies that care about safety and security are always looking for well-trained employees. That presents an opportunity for academic institutions to foster strong safety and security practices and build relationships with industry to ensure employment of their students, and enhance the culture of safety and security compliance within academic laboratories.

In addition, industry and trade organizations in developing nations (such as the Chemical Industries Association of the Philippines and the Plastics and Chemicals Industries in Australia) can help the CSP coordinate with universities and professional societies in their regions to harmonize safety and security controls.

## Certification and Training Programs

The CSP may wish to take advantage of existing training programs for chemical safety professionals, such as the following:

- The Certified Safety Professional Program of the Board of Certified Safety Professionals (BCSP) in Savoy, Illinois, is designed to recognize the qualifications and experience of safety professionals.[26] Applicants must apply to BCSP and be approved to sit for examinations. Requirements include a college degree (bachelor's in any field or associate's in a safety-related field) and professional employment in which the primary function is safety

[25] See the European Chemicals Agency Web site for more information about implementation of REACH: *http://echa.europa.eu/home_en.asp* (accessed January 22, 2010).

[26] For more information, see *www.bcsp.org* (accessed October 26, 2009).

(prevention of harm to humans and the environment). Examinations are offered in the United States and abroad.

- The National Registry of Certified Chemists-Certified Chemical Hygiene Officer (CHO) program is designed primarily for those who are responsible for laboratory safety in their institutions or organizations.[27] Because U.S. regulations require the appointment of a CHO for a laboratory facility to comply with requirements of the Occupational Safety and Health Administration,[28] the program is designed largely for U.S. laboratory personnel; however, it can be taken by others. Requirements include education and experience in chemical safety.
- The American Board of Industrial Hygiene-Certified Industrial Hygienist (CIH) Program is designed for professionals who are involved in ensuring the health and well-being of workers and their communities. Typically, they are involved in evaluating the health effects of chemicals in a workplace or community. Most CIHs are in the United States and Canada. The certification is focused on the science and art of recognition, evaluation, and control of safety, health, and environmental factors in the workplace that may result in injury, illness, or impairment, or otherwise affect the well-being of workers and members of the surrounding community. The technical knowledge tested includes: basic sciences; occupational diseases, toxicology, and health hazards; and work environments, principles of investigation methods, ethics, risk communication, guidelines, and standards. CIH certification also means that the professionals have skills in data management and integration, emergency response, hazard evaluation (instrumentation, sampling methods and techniques, and analytical chemistry), and hazard control (engineering, personal protective equipment, and administration).

## International Standards

The ISO (International Organization for Standardization)[29] develops and publishes international standards on scientific and technical issues in industry. Many institutions in developing countries are increasingly seeking to attain certification in international laboratory standards set by ISO, such as the Quality Management Systems standard (ISO 9001) and the general requirements for the competence of testing and calibration laboratories

[27]For more information, see the NRCC website at *www.nrcc6.org* (accessed October 26, 2009).

[28]Occupational Exposure to Hazardous Chemicals in Laboratories. 29 CFR 1910.1450. *http://www.osha.gov/pls/oshaweb/owadisp.show_document?p_table=standards&p_id=10106* (accessed January 22, 2010).

[29]For more information, see *http://www.iso.org/iso/home.htm* (accessed October 26, 2009).

(ISO 17025). ISO 9001, for example, sets out eight principles that help guide institutional policies and practices:

Principle 1: Customer focus;
Principle 2: Leadership;
Principle 3: Involvement of people;
Principle 4: Process approach;
Principle 5: System approach to management;
Principle 6: Continual improvement;
Principle 7: Factual approach to decision making; and
Principle 8: Mutually beneficial supplier relationships.

Security is not addressed in any of the standards focused on chemical laboratory operations. The CSP and partnering organizations, such as IUPAC and ACS, should press the ISO to develop new standards that address chemical security, which could raise awareness and drive adoption of chemical laboratory security practices in developing countries.

# 3

# Guidelines for Establishing a Chemical Laboratory Safety and Security Program

Establishing a culture of safety requires sustained commitment to high standards at all levels—from top institutional leaders to laboratory personnel. As developing countries establish safety and security programs, it is important for them to consider the entire life cycle of chemicals—from planning, procurement, and security to ultimate use and final disposal. This chapter recommends a framework for systematically integrating safety and security into a research institution to anticipate and prevent circumstances that might result in injury, illness, or adverse environmental effects. The way the individual elements of the framework are applied will depend on the size of the institution, the nature of its activities, and the hazards and conditions specific to its operations. Chapter 4 provides further guidance in enhancing compliance with safety and security policies, programs, and rules.

## WHOSE JOB IS IT? RESPONSIBILITY FOR CHEMICAL LABORATORY SAFETY AND SECURITY

Individuals in an institution have various roles and responsibilities for establishing and maintaining safe and secure practices. Setting a good example is the best way for people at all levels to demonstrate their commitment.

### Leaders

Responsibility for safety and security rests ultimately with the head of the institution and its operating units. In some cases, there may be legal obligations and large personal fines or prison sentences if they do not provide a secure and safe working environment. Leadership by those in charge ensures that an effective safety and security program is embraced by all; even a well-conceived program will be treated casually by workers if it is neglected by top management.

### Safety Officers

To establish and support a unified effort for safety management and to provide guidance to people at all levels, each institution should have at least one designated safety officer. The safety officer should be equipped with the knowledge, responsibility, and authority to develop and enforce an effective safety and security management system.

### Environmental Health and Safety Office

Some larger institutions also have an environmental health and safety office staffed by experts in chemical safety, engineering, occupational medicine, fire safety, toxicology, and other fields. Such an office assists in establishing policies and promoting laboratory safety standards, and often handles hazardous waste issues, accident reviews, inspections and audits, compliance monitoring, training, recordkeeping, and emergency response.

### Chemical Laboratory Managers and Instructors

Direct responsibility for the management of a chemical laboratory safety program typically rests with the chemical safety officer (CSO) and a laboratory manager. In coursework, laboratory instructors carry direct responsibility for actions taken by students. Instructors are responsible for promoting a culture of safety and for teaching the skills that students and other workers need if they are to handle chemicals safely.

### Students and Workers

Although they are influenced by and depend on attitudes of and guidance by those in leadership or management positions, students and other laboratory workers who actually do the work are responsible for working safely.

## 10 STEPS TO ESTABLISH AN EFFECTIVE CHEMICAL LABORATORY SAFETY AND SECURITY PROGRAM

The most critical aspect of establishing a strong safety and security program is the commitment and support it should have from the top leaders in the institution. If the leaders facilitate the establishment of this program and hold their managers responsible and accountable, a culture of compliance can be achieved.

**1. Establish an institutional safety and security oversight committee and designate a chemical safety officer.** The top institutional leader (such as the president or chancellor of a university or the director of a research center or agency) should establish a committee to provide oversight for chemical safety and security. The committee should report directly to the top leaders and receive the necessary financial and administrative support. The institution should also define CSO responsibilities and authority and appoint at least one CSO. Each CSO should be part of the Safety and Security Oversight Committee.

For a CSO to be effective, he or she will need to be given dedicated time, resources, and the necessary authority to carry out his or her responsibilities. CSOs should have direct access, when necessary, to the senior authorities who are ultimately accountable to the public.

**2. Develop a chemical safety and security policy.** Institutional leaders should implement a formal policy to define, document, and endorse the program, and the policy should integrate safety and security into the entire life cycle of all laboratory chemicals. A formal policy statement establishes expectations and communicates the institution's intent. The policy should state how the institution will

- prevent or mitigate human and economic losses arising from accidents, adverse occupational exposures, and environmental events;
- build safety and security considerations into all phases of operations;
- achieve and maintain compliance with laws and regulations; and
- improve performance continually.

The policy statement should be communicated and made readily accessible to all employees and should be reviewed and revised by top management as often as necessary. An example of a policy statement can be found in Appendix E. In developing a chemical safety and security policy, laboratory management should establish a credible and strong compliance program, complete with accountability and consequences for noncompliance (see Chapter 4). Management should also communicate a commitment to safety and security to all laboratory personnel regularly.

**3. Implement administrative controls and processes for performance measurement.** Administrative controls are an institution's specific rules and procedures for safe and secure practices, and they establish the responsibilities of the personnel involved.

CSOs should develop general safety rules, laboratory housekeeping procedures, manuals for use of materials and equipment, and other documents to communicate rules and expectations to all laboratory personnel. Those documents should clearly define the individual responsibilities of students and other laboratory workers, laboratory managers, institutional leaders, contractors, emergency service providers, and visitors to the laboratory.

The business of conducting experiments changes continually: attrition in staff members and students, change in regulations, and evolution of technologies. As a result, evaluating the safety and security aspects of chemistry laboratory operations should be part of everyday activities, in addition to and separate from regular formal performance measurements and reviews. For example, beginning all department and group meetings with a safety moment is one way to highlight the importance of safety. Administrative controls should provide mechanisms for managing and responding to change, such as new procedures, technologies, legal requirements, staff, and institutional changes.

In addition to performance measurements by those working in the laboratory, periodic audits by knowledgeable people independent of the location or activity should be arranged to enable a deeper and more critical appraisal. Top management should review the management system and its formal policy regularly.

**4. Identify and address particularly hazardous situations.** Managers, principal investigators, lead researchers, team leaders, and supervisors should take active roles in managing the safety and security of their laboratories. An initial status review to assess the scope, adequacy, and implementation of safety procedures will provide a foundation on which to build a robust safety and security program and will help in setting priorities among efforts to improve safety and security. A risk-based evaluation should be performed to determine the adequacy of existing control measures, to set priorities among needs, and to incorporate corrective actions according to importance and available resources.

To begin the process of ensuring effective management of chemicals, laboratory management should establish a list of all the chemicals in the laboratory, especially the chemicals of concern (COCs). COCs are highly hazardous chemicals or chemicals that are potential precursors of highly hazardous materials. Typically, the list would include chemicals listed by the Chemical Weapons Convention, chemicals that have potential for mass destruction, explosives and precursors of improvised explosive devices, and chemicals of high acute toxicity (rated as Category 1 in the Globally Harmonized System of Classification and Labeling of Chemicals). Examples of COCs are provided in Appendix D.

**5. Evaluate facilities and address weaknesses.** The role of physical access control in improving security of chemicals, equipment, and occupants of buildings in which chemicals are stored and used should be addressed specifically. This will require development of a comprehensive security vulnerability assessment and policy setting (see "Guidelines for Facility Access and Use" later in this chapter).

**6. Establish procedures for chemical handling and management.** Chemical management is a critical component of a laboratory program. Safety and security should be integrated into the entire life cycle of a chemical, including procurement, storage and inventory, use and handling, and transport and disposal. The overall process is described in more detail later in this chapter (see "Procedures for Managing and Working with Chemicals of Concern"). It should include procedures for screening of COCs as part of the normal procurement process. An inventory process to track use of a chemical until it is completely consumed or finally disposed of should be established. For example, when chemicals are received, their identities and quantities are entered into the inventory system. The inventory and record-keeping system can be important in order to

- ensure the security of chemicals through accountability of chemical usage;
- provide a resource to consult for possible sharing of chemicals;
- provide information that allows managers to know when to reorder chemicals;
- provide the location of hazards in the laboratories for emergency responders;
- determine future needs and uses of chemicals; and
- minimize excess inventory and chemical waste.

An important part of any process is accountability for chemical use and adherence to procedures. Managers should consider ways to recognize and reward those who follow best practices while handling and working with chemicals. Alternatively, managers may need to consider tools for enforcement of the practices when investigators bypass the system.

**7. Use personal protective equipment and engineering controls.** Engineering measures, such as a laboratory hood, local exhaust ventilation, and a glove box are the primary methods for controlling hazards in the chemical laboratory. Personal protective equipment (PPE) such as safety glasses, goggles, and face shields are used to supplement engineering controls. Laboratory management should not allow an experiment to proceed if inadequate hazard control measures (engineering or PPE) are unavailable. More detailed guidance on procedures for chemical handling and management are provided in the section "Special Concerns" later in this chapter.

**8. Plan for emergencies.** Laboratories, like all other workplaces, experience unplanned incidents and emergencies. Laboratories should make plans to handle emergencies and implement the plans by purchasing and maintaining emergency equipment and supplies, such as fire extinguishers, eye washes, safety showers, and spill kits. The use of a COC may warrant development of special plans, such as antidotes for unintentional exposures (for example, atropine for organophosphorous agents). Some COCs may be pyrophoric (ignite spontaneously) and require special fire-extinguishing methods. Emergency response preparedness should involve local emergency response organizations, such as fire departments, to ensure that they have the equipment to assist in the event of an emergency. *Prudent Practices in the Laboratory*[1] or other safety manual should be referred to for more information.

**9. Identify and address barriers to safety compliance.** Compliance with good safety and security practices involves having people act in accordance with established institutional policies and procedures. Each country faces challenges in implementing effective safety and security practices and complying with them. Local culture often presents barriers to compliance, and efforts are needed to address and overcome the barriers, as discussed in detail in Chapter 4.

**10. Train, communicate, mentor.** A comprehensive process that manages the entire life cycle of a chemical in the laboratory would result in responsible management of the safety and security aspects of that chemical. The CSO is responsible for ensuring that proper processes are established and communicated to all, but it takes a strong commitment by top leaders in the institution to create the best safety and security systems and to establish a culture that ensures the well-being of personnel and the public.

---

[1]National Research Council. *Prudent Practices in the Laboratory: Handling and Management of Chemical Hazards, Revised Edition.* Washington, D.C.: The National Academies Press, in press.

Accordingly, top leaders are ultimately accountable for chemical safety and security. See the section on "Guidance on Assigning Responsibility and Accountability" for more information.

## SPECIAL CONCERNS

### Security of Chemicals

Chemists and other scientists collectively use thousands of chemicals in their laboratory work, but COCs pose a particular risk to the general public if they are acquired by people who wish to inflict harm. Some COCs can be used as precursors to make potentially deadly chemical agents or illicit drugs. Other chemicals have hazardous properties that can pose risks to laboratory personnel, especially if they are unaware of the properties. The United Nations has developed a system, the Globally Harmonized System for Classification and Labeling of Chemicals, for classifying chemicals according to their hazardous properties. In that system the most hazardous chemicals make up Hazard Class 1; these COCs should be kept secure from theft or diversion, and there are laboratory security measures that can help to prevent such diversions.

Security begins with the individual, and prudent security practices for laboratories should include a system that limits access to authorized personnel who have a need to work in the laboratory. Authorized personnel should be approved to work in a particular area and have access and authority to use COCs by someone in the institution. Authorized personnel should be given access through keys or card keys and may have identification badges. The issuance of keys or card keys should be subject to an established process in which laboratory personnel sign for keys and turn them in when they leave the program. Keys should be of a type that is not readily duplicated.

In the academic community, security measures can pose a challenge. Teaching laboratories generally have few types and small amounts of chemicals, and the chemicals are not likely to be COCs; however, many teaching laboratories might have a wide range of solvents, some of which are COCs. In general, more hazardous chemicals are used in advanced laboratories, especially research laboratories, and there should be greater attention to limiting access to those laboratories. Separating advanced research laboratories from other laboratories will make security much easier. In commercial and government institutions in some countries, it is common to conduct background checks of personnel who will work in research laboratories and have access to COCs. It is not as common in academic institutions, particularly for students. It takes vigilance on the part of those who work in the laboratories to look for suspicious activity or unexplained missing COCs or to detect security breaches.

Physical security is an important part of any security program. When given proper instructions and training, security guards can play a key role in preventing access to areas where COCs are used or stored. Door locks, of either the normal key type or the more expensive electronic type, are necessary where COCs are being used or stored. Other physical security measures include locked cabinets, locked storage areas, locked drawers, and perhaps alarm systems. Any extraordinary laboratory security measures should be commensurate with the potential risks and should be imposed in a manner that does not hamper research or safety unreasonably.

The effectiveness of a security program is closely related to the expectations established by management. All laboratory personnel should be encouraged to question the presence of unfamiliar people in laboratories and to report suspicious activity immediately. Locking laboratory doors when laboratories are unoccupied should become routine. COCs that are not being used should be secured. Laboratory managers should establish a policy that prohibits all unauthorized use of laboratory materials and facilities, and that violations will be subject to a penalty.

A training program in laboratory security should set out the expectations of management and the need to maintain a safe and secure laboratory environment. Training should be conducted periodically and especially for new personnel. Laboratories should be inspected routinely for compliance with security measures. Personnel working in laboratories should follow all established security procedures, and there should be a protocol for reporting security breaches or security concerns.

## GUIDELINES FOR FACILITY ACCESS AND USE

Conducting a security vulnerability assessment (SVA) and developing and writing a site security plan as described below are meant to be in conjunction with establishing an effective safety and security program (see Step 5 above).[2]

---

[2]Many of the ideas for this section are based on the National Institute of Justice document *A Method to Assess the Vulnerability of U.S. Chemical Facilities* (November 2002). Other documents that were useful are the state of South Carolina's *Best Practices: Workplace Security* (February 2003), the Occupational Safety and Health Agency's *Compliance Policy for Emergency Action Plans and Fire Prevention Plans* (July 2002), the National Institute for Occupational Safety and Health's *Guidance for Protecting Building Environments from Airborne Chemical, Biological, or Radiological Attacks* (May 2002), and the U.S. Department of Justice's *Using Crime Prevention Through Environmental Design in Problem-Solving,* by Diane Zahm (August 2007). Organizations identified in Chapter 2 may offer further resources.

## Developing a Comprehensive Security Vulnerability Assessment

An SVA may include an entire campus or specific facilities on a campus and it involves a series of comprehensive investigations and an integrated analysis. The purpose of an SVA is to catalog potential security risks to a laboratory, determine the magnitude of the risks, and assess the adequacy of systems that are in place. An SVA helps in determining the security planning needs of a facility. An SVA should include an asset evaluation, threat assessment, site survey and analysis, and physical vulnerability survey.

### Asset Evaluation

This investigation identifies and quantifies valuable assets—such as equipment, instruments, libraries, and documents—that should be protected from accidental loss or damage and from theft or destruction by persons who intend to do harm or by natural disasters. Information should be included about sources of replacement and alternative resources on campus or elsewhere that could permit continuity of operations.

### Threat Assessment

This identifies possible types of threats to the institution and specific facilities from the generic to site-specific threats, from natural disasters to terrorist attacks. To the extent possible, a threat assessment should describe the adversarial groups or individuals, their ideological and economic motivations, members and supporters, leadership and organizational characteristics, record of illegal or disruptive activities, preferred mode of action, and potential capabilities to attack a target, what they typically want to communicate to the public, and how they prefer to do it. Institutions must to be careful to adhere to laws that protect personal privacy within their country. Possibilities of attack or action against the institution and its facilities should be detailed. The consequences of natural disasters—including wind, water, fire, earthquake, and multifocal events such as those that occur during cyclones, hurricanes, tornados, earthquakes, tsunamis, and volcanic eruptions—should be estimated. Scenarios (best case and worst case) should be generated to derive a measure of the potential severity of an event, natural or malicious. Chapter 4 of *Prudent Practices in the Laboratory* and the American Chemical Society's SVA provide more complete explanations of this process.

### Site Survey and Analysis

This part of an SVA is specific to the physical facilities covered by the security and facility access policy. Up-to-date drawings of campus features, vehicular traffic, pedestrian circulation, site and terrain, and buildings are critical resources for this investigation. Walk-throughs of specific buildings that use or store chemicals, and a tour of the entire site or campus, should be conducted. These inspections should be documented with photographs or videos of specific conditions.

Building enclosure integrity with regard to weather and physical intrusion are important to investigate in all areas, on all sides, and on roof and subsurface extensions, including tunnels and utility routes and entry points into buildings. Locations of air intakes for mechanical and natural ventilation and locations and conditions of storage elements for chemicals and other hazardous materials are important to analyze.

The site survey and analysis should include a vehicular traffic plan that highlights areas for material deliveries, truck routes, parking, and building entries and exits. The site analysis should address traffic patterns of vehicles and pedestrians over 24-hour periods on normal workdays and weekends, physical protection and security features, building uses, and which persons are allowed access. Such a comprehensive review is necessary to permit an accurate survey of physical vulnerability and to put into place operational procedures for detection, delay, and assessment systems to protect physical assets and to protect operations on the campus or in a facility that could be interrupted or sabotaged.

### Physical Vulnerability Survey

A vulnerability survey includes several kinds of investigation, within the limits of local legal frameworks.

1. Identifying potential targets and gaining access to those targets.
2. Identifying and rating potential threat(s) based on historical context; for example, threats that have previously been acted upon have more significance than potential threats without historical motive or intent. This pertains to both natural factors, such as the likelihood of flood, and malicious action.
3. Identifying personnel, contract employees, vendors, contractors, and visitors who may have personal problems or conflicts with the institution and who may also be able to identify internal physical facility vulnerabilities and obtain access to facilities.

Various data should be considered in a vulnerability survey.

- Which potential targets are clearly recognizable with little or no knowledge.
- Which potential facility targets store chemicals.
- The quantities, concentrations, and hazards of the chemicals that could be involved in each potential target.
- The potential for offsite release or illegal use of the chemicals.
- Physical protection measures that are in place to mitigate the harm that could result from a chemical release or spill.

A matrix or other analytical tool should be devised to estimate the severity of effects of each undesired event identified in the scenarios developed in the threat analysis. The severity level will contribute to the overall risk analysis. Worst-case scenarios should be used for events (natural or malicious) for estimating

- how many people would be affected;
- what the monetary loss of property would be;
- how much money and time would be needed to acquire replacement facilities;
- what the loss of productivity and the period of shut-down and recovery would be; and
- what value in public trust, support, and image would be lost.

## Develop a Site Security Plan

A comprehensive site security plan integrates all the information gained in the analyses, surveys, and investigations mentioned above. It addresses workplace security guidelines and an emergency response plan that provides a physical protection strategy to detect, delay, and respond quickly and effectively to interrupt, neutralize, or mitigate malicious-intent threats and natural disasters. Methods in the public domain (such as Responsible Care, CEFIC, IUPAC, and ISO) outline many approaches to developing a plan that meets the goals of the security and access control policy. Tactics and a wide variety of technologies are well documented to provide in-depth protection and to minimize consequences of failure of security components.

Institutions may consider applying concepts of crime prevention through environmental design. Building location and properly placed entryways, windows, lighting, shrubbery, and other physical features can help deter criminal activity, and are cost-effective, systemic improvements that do not depend solely on technology. The security and access control policy should be the basis of the site security plan.

## PROCEDURES FOR MANAGING AND WORKING WITH CHEMICALS OF CONCERN

### Procuring Laboratory Chemicals

Laboratory management should establish a system for authorizing procurement of a COC that considers all the acceptable ways that COCs may be obtained in laboratories—such as purchase order, credit card order, online purchase, donation, sharing, exchanging. The system should require that the CSO be notified of orders of COCs and, if necessary (for example, when there is an unusual supplier or material that has not been purchased previously), the CSO's approval should be obtained. The CSO should be notified when the chemical is received, and the CSO or a designated alternate should check the order and enter the amount in an accountability log. The CSO or designated alternate should maintain an inventory accountability log for COCs, from procurement to disposal. The log should be kept in a secure location where everyone can read it but only the CSO or a designated alternate can edit it. If a centralized log for the department is not adequate, each laboratory or group should keep its own log. The log could be an automated system, spreadsheet, or a manual list of all COCs in the laboratory. The important aspects are that it be diligently maintained and that compliance be enforced; otherwise the effort will be wasted. The log should identify each chemical to avoid ambiguity (for example, by using the Chemical Abstract Service registry number) and should indicate use and current inventory. The CSO or designated alternate should review the logs regularly, for example, monthly. An example of an inventory accountability log is provided in Appendix F.

### Storing Laboratory Chemicals

Chemicals should be stored safely and securely on the basis of risks and hazards, as outlined in *Prudent Practices in the Laboratory*. Access to the storage area should be limited to persons approved by the laboratory manager or supervisor and the CSO. The list of authorized persons should be posted at the storage facility and communicated to all laboratory personnel. It is important to maintain an appropriate level of security (for example, door locks, lock boxes) for all chemicals, especially COCs (as discussed earlier). Laboratory managers should remind users that COCs should be secured when not in use. There should be a requirement that unwanted or unneeded chemicals are returned to the storage area at the end of a project or process. Experience indicates that the CSO needs to reinforce this requirement regularly.

**BOX 3-1**
**Five Questions for Laboratory Safety and Security**

1. What are the hazards? For example, health hazards, flammability, reactivity, and physical hazards.
2. What is the worst thing that could happen? For example, personal exposure, spills, fire, uncontrolled reaction, and electric shock.
3. What can be done to prevent this from happening? For example, substitution, guarding, change in environmental conditions, and modification of a procedure.
4. What can be done to protect from these hazards? For example, ventilation, gloves, eye and face protection, and protective clothing.
5. What should be done if something goes wrong? For example, spill control, fire extinguisher, safety shower, and eyewash.

### Safely Using Laboratory Chemicals

There are established ways to manage the safe use of COCs and other hazardous chemicals. The critical elements of safe work with COCs described below are known to most health and safety professionals. Box 3-1 provides a useful checklist for use of any hazardous chemical. It can be posted in the laboratory or given to laboratory workers. The first step in ensuring safety is to conduct hazard evaluations to understand the nature and severity of the risks being posed by the chemicals being used. Once the risks are determined, the best available measures to control and manage them should be selected. Those measures include developing a specific safety plan or safe operating procedures (SOPs) for the particular intended use of the chemicals. It is critical to ensure that all who will handle chemicals are knowledgeable about the hazards and safe handling procedures, especially COCs. It is equally important to ensure that users follow established safety procedures during their work. Hence, communication of the safety procedures to all users is essential, and a suitable process for that should be developed. Users of extremely hazardous chemicals may require additional training.

Unexpected incidents or events during handling of COCs or other hazardous materials are almost inevitable, so a system that requires and encourages the reporting of all incidents, concerns, or problems (even if minor) should be established. Such a system allows reviews and continuous improvement of the processes and systems that are in place.

## Hazard Evaluation

The risks posed by use of a chemical can depend on the quantity used. The larger the quantity of a COC being used, the more serious the hazard; reducing the quantity being used (or stored) reduces the hazard. The CSO or designated alternate should evaluate the hazards posed by the chemicals in the quantities procured and used. The hazard evaluation should consider the routes of potential exposure: eye contact, skin contact, inhalation, injection, and ingestion (can be substantially reduced by prohibiting eating and drinking in the laboratory). It should take into account physical hazards and health hazards. Physical hazards can include flammable or explosive chemicals, high-vacuum or high-pressure systems, hot equipment or devices, cryogenic materials, radioactive materials, and corrosive acids or bases. Combinations of chemicals that result in extreme exothermic (heat-releasing) reactions can result in explosions if not properly controlled. Health hazards include acutely toxic chemicals, sensitizers (allergens), chemicals that cause chronic toxicity (such as carcinogens), and reproductive toxicants. Particular handling practices and procedures should be developed for laboratory hazards (see the hazard assessment checklist in Appendix F).

## Hazard Control and Management

In all laboratory experiments and procedures, it is important that hazards be controlled or managed primarily by engineering measures, such as a laboratory hood, local exhaust ventilation, or a glove box. Ventilated enclosures can often be used for weighing chemicals. Best practices also include having a laboratory under negative pressure with respect to the adjacent hallway so that hazardous chemical vapors are kept in the laboratory. In same cases, such as with the use of radioactive materials, personal hand, foot, or full-body monitors may be needed to control the spread of material into areas outside controlled laboratory facilities.

Another method of controlling and managing hazardous chemical operations could be the use of smaller-scale experiments or procedures that use micro-scale or mini-scale laboratory equipment. That reduces the quantities of chemicals and thus reduces risk.

Although the primary methods of control are engineering methods, personal protective equipment (PPE) should supplement them. Eye and face protection (safety glasses, goggles, and face shields) are essential in any laboratory in which chemicals are handled. Protective gloves are important if chemicals can come into contact with skin. Gloves should be selected for the particular chemicals being used, recognizing that no single kind of glove protects against all hazards. Laboratory coats and aprons provide additional protection for the body from chemical exposure. Protective footwear may be needed in some laboratories.

Respirators are a last resort for controlling a hazard and preventing exposure. They should not be used unless all other methods of control are inadequate. Respirators should be properly selected, should be correctly fitted to a person's face, and a person using a respirator should be aware of its uses and limitations. It is essential that all respirator users receive training in respiratory protection and that respirator users must work in teams of at least two workers.[3]

If the hazard control measures (engineering or PPE) are not adequate to prevent exposure and provide an acceptable level of safety, laboratory management should not permit the experiment or procedure in question, and should explore whether a safe substitute for COCs can be identified.

### Safety Plans for Chemical Use

Safety plans or SOPs should be developed to document the hazards posed by and controls for COCs. SOPs can have a variety of formats, including postings, documentation in laboratory notebooks, forms, binders, and digital collections of plans. The objective is to document hazards and control and disposal methods in a manner that ensures that those who have access to the materials will understand how to use them safely. An example of a form for a safety plan is provided in Appendix F.

### Incident Reporting

A process for reporting and investigating incidents should be established. It should emphasize free exchange of information without penalty to the people who report an incident. The objective is to (1) maintain a culture in which people feel comfortable sharing information about problems they have encountered and about their concerns and (2) promote the understanding that laboratory workers' personal safety is paramount.

## Disposing of Laboratory Chemicals

The end of the life cycle of a chemical is its consumption in a laboratory procedure or its disposal. The CSO or designated alternate should develop a program to ensure safe and environmentally responsible disposal of chemicals, especially COCs.

---

[3]For more information, see the OSHA Respiratory Protection Standard: *http://www.osha.gov/pls/oshaweb/owadisp.show_document?p_id=12716&p_table=standards* (accessed January 26, 2010).

Before procuring chemicals, experimenters should ensure that disposal facilities that will manage them are available or that there is a laboratory SOP for rendering them safe for drain or solid-waste disposal. Procedures for handling waste from the time it is produced to its ultimate treatment or disposal need to be developed. Laboratory managers should ensure that laboratory workers know how to collect waste safely and the type of containers to use. Laboratory workers should be given guidance on how, how much, and where to store waste in the laboratory as it is collected, including methods to prevent spills and accidental releases. Keeping containers sealed when they are not in use and using secondary containment, such as trays or bins, in the event of a spill, are best practices.

Procedures for removing chemical waste from the laboratory for storage in a facility or area dedicated for that purpose or for removal by a qualified chemical waste vender need to be formulated. A process whereby laboratory managers notify the CSO or waste manager that they have chemical waste for removal should be established.

Depending on the location of the facility, there may be regulations or guidance on recordkeeping for chemical waste. Records of disposal—including disposal date, quantity disposed of, and disposal method—should be maintained. The records should be kept indefinitely or as long as specified in regulations.

## Guidance on Assigning Responsibility and Accountability

The following are examples of how responsibility and accountability for chemicals may be assigned. The needs and management style of an institution should be considered in determining its appropriate management structure.

### Responsibilities of Chemical Safety and Security Oversight Committees

The Chemical Safety and Security Oversight Committee, appointed by the top leader of the institution, is responsible for the following broad elements:

- developing and maintaining safety and security policy initiatives;
- adequately budgeting and allocating resources for the chemical safety and security program;
- making inquiries as appropriate regarding incidents, accidents, and breaches of safety and security;
- making recommendations to top leaders regarding recognition of best practices and disciplinary actions as appropriate; and
- providing necessary support to the chemical safety officers (CSOs).

**Responsibilities of Chemical Safety Officers**

CSOs are responsible for the following broad elements:

- developing and implementing an integrated safety and security program for the life cycle of the chemicals of concern (COCs) on the basis of the guidelines established in the present report and discussed in more detail in *Prudent Practices in the Laboratory*;
- establishing a training program that ensures that laboratory managers, supervisors, and workers receive training appropriate for their duties and for the materials they will use regularly, reviewing and approving the content of the training, and maintaining training records and periodically auditing them to ensure that all who should be trained are being trained;
- developing a laboratory inspection program that reviews laboratory facilities, SOPs, and worker preparedness; keeping records of the inspections and sharing the results with laboratory managers and senior management; and tracking resolution of issues identified during inspections;
- periodically auditing all aspects of the safety and security program and reporting findings to senior management, including recommendations for improvements; and
- developing a program for managing incidents, including spills, injuries, and near misses.

More specifically, CSOs or designated alternates are responsible for

- executing the established policies regarding laboratory COCs and ensuring compliance with applicable regulations as required;
- assisting in procurement, storage, use, and waste disposal at the laboratory level, including providing training for how to develop appropriate SOPs;
- operating, if required, a waste management program for wastes to be disposed of outside the laboratory, including receipt of wastes, transportation, and final disposal of material by commercial venders;
- logging orders of COCs;
- receiving and inventorying on purchase receipts;
- disposing of laboratory COCs;
- auditing inventory logs and cabinet security at least once a year;
- investigating incidents involving COCs;
- suspending authorizations to use laboratory COCs in cases of noncompliance; and
- maintaining complete records of program operations in a form suitable for inspection that can readily be retrieved and distributed.

### Responsibilities of Laboratory Managers or Supervisors

Each laboratory should designate a manager or supervisor. In many cases, the person may be the principal investigator. Laboratory managers and supervisors play an important role in the safety and security program and will be the key connection between the laboratory and the CSO. Responsibilities include at least

- ensuring that laboratory workers receive training in general chemical safety and security;
- ensuring that laboratory workers understand how to work with COCs safely and providing chemical-specific and procedure-specific training as needed, including developing and reviewing SOPs;
- providing laboratory workers with appropriate engineering controls and PPE needed to work safely with COCs;
- ensuring that the laboratory has the appropriate level of security for COCs;
- setting expectations for safety and security and including safety and security components in performance appraisals; and
- reviewing and approving work with COCs.

### Responsibilities of Local Laboratory Safety Committees

Local laboratory safety committees are composed of managers of laboratories, laboratory workers (or their selected representatives), CSOs, and perhaps security officers. Responsibilities include

- reviewing and discussing current safety issues;
- reviewing safety incidents;
- planning safety training;
- providing representation for laboratory workers; and
- making recommendations for safety improvements.

### Responsibility of Laboratory Workers

All the people who work in a laboratory—whether paid or unpaid, student or employee—are responsible for following all the safety and security protocols for their own protection and the protection of their fellow workers. All laboratory workers are responsible for at least the following:

- attending laboratory safety training;
- reviewing and following written procedures;

- ensuring, before working with a chemical or procedure the first time, that all the hazards and procedures needed for safety and security are understood by either reviewing or developing and approving SOPs;
- asking a supervisor or a CSO for help if unsure about the hazards;
- using engineering controls and PPE, as appropriate;
- reporting all incidents, security issues, and potential chemical exposures to a laboratory supervisor; and
- documenting specific operating procedures for working with COCs and amending procedures as needed.

# 4

# Compliance with Safety and Security Rules, Programs, and Policies

Establishing rules, programs, and policies for laboratory safety and security is of no value if organizational leaders do not enforce them and if laboratory managers and workers do not follow them. Incentives are needed to ensure that laboratories operate safely and securely and comply with established organization rules, programs, and policies. In its outreach to top institutional leaders, the U.S. Department of State Chemical Security Engagement Program (CSP) should encourage institutions to develop a system of compliance with safety and security rules, programs, and policies. Organizations also need to identify the barriers to chemical laboratory safety and security in their cultures and find ways to overcome them. This chapter recommends a system for addressing the barriers to and fostering compliance with good laboratory safety and security practices.

## COMPONENTS OF A GOOD COMPLIANCE SYSTEM

The major components of a compliance system are regular inspections, reporting, incident investigation, follow-up, enforcement, and recognition and reward. The system should emphasize fact finding, not fault finding. That applies to all the safety and security programs and policies described in Chapter 3. Initiation and maintenance of an effective compliance system are important to:

- give organization leaders useful information about the effectiveness of safety and security systems and about needs for improvements;

- give designated safety and security personnel authority to collect incident reports and report incidents to higher authorities for action;
- discern patterns of unsafe behavior and facilities (based on statistics from reports and inspections), find methods to improve safety and security, and initiates new rules and regulations to protect workers and students;
- increase awareness of safety issues in the organization so that a culture of improved safety and security is encouraged;
- give current information to safety officers so that training of all laboratory workers can be improved and specific guidance can be given to individual workers; and
- give information to laboratory leaders so that they can learn how to use, test, and procure appropriate personal protective equipment (PPE) and other types of equipment to improve safety.

## Inspections

There should be a program for regular inspections of all science and engineering, safety and security practices, and facilities. Conducting an inspection is just the first step; issues found should be resolved to achieve a safer and more secure status. Written communication and documentation of inspections and of resolution issues are essential (see Appendix G for sample inspection checklist).

Conducting inspections also gives chemical safety officers (CSOs) opportunities to notice and reward best practices and to communicate them to the larger scientific community. Leaders of the organization may want to authorize CSOs to recommend individuals or groups for special recognition and even material reward.

## Reporting

A process for incident reporting and investigation should be established, with an emphasis on free exchange of information without penalty to the persons who report an incident (see Appendix G for a sample incident report form). The objectives are to maintain a culture in which people feel comfortable in sharing information about problems they have encountered and promote an understanding that laboratory workers' personal safety is paramount.

Complex hierarchical systems in developing countries sometimes suppress individual responsibility. An organizational support system and a fundamental change in the behavior of individuals are essential to enable effective reporting of accidents, incidents, and lapses. The greatest challenge is to reduce resistance to reporting problems.

Scientific leaders and administrators should regard the reporting system as a method of furthering education and training of valuable skilled workers and students, not as a means of justifying punitive actions. That requires a fundamental cultural change in organizations to conduct bold and open discussion among employees, students, and leaders. Credibility is established by actions. If an organization's leaders use accident or incident reports as the basis of punitive actions against particular employees, the reporting system will never take root and foster the culture of safety.

Building a culture of safety involves not only increasing recognition of specific potential hazards but also helping workers and students to make better, safer choices in their actions. They should have confidence in the fairness and objectivity of their organization's leaders.

## Enforcement

Both positive and negative feedback is necessary to ensure the proper enforcement of safety and security rules and regulations. The reporting system should delineate consequences of not reporting incidents and not complying with safety and security rules. Establishing rewards for individuals and groups that display consistently safe behavior would reinforce the desired behavior. Workers and students should be encouraged to speak up when they witness incidents, lapses in abiding by safety rules, or outright violations. Such laboratory incidents as sink fires, chemical-hood fires, chemical spills, waste disposal accidents, and safety shower activations need to be reported to a CSO and the laboratory supervisor. They should not be considered trivial even if there is no immediate consequence, such as a call to a fire department or a trip to a hospital emergency room.

### Safety Incidents

Laboratory supervisors are responsible for reporting safety incidents in their laboratories. A form should be filled out that indicates clearly the name of the person involved, the name of the department, the date and time of the incident, and details of the factors that contributed to it. Penalties for not reporting should be severe enough to discourage hiding safety incidents.

### Security Breaches

All security breaches, small or large, need to be reported in writing, to the concerned authorities. That requires an atmosphere of openness and confidence in the rules and in the leaders. Reporting security breaches helps to improve security systems. People who report security breaches immediately should be rewarded.

### Suspicious Activity

All personnel should be trained to look out for suspicious activities or persons. They should learn to report such activities in a timely manner. Persons who do so should receive special recognition from organizational leaders.

## Best Practices

The laboratory community should be encouraged to report outcomes of inspections. As mentioned earlier, positive recognition of good practices during an inspection constitutes effective encouragement of a culture of safety.

### Protection for Those Who Report Incidents

Clearly written rules should be established to protect those who witness and report a safety or security incident or suspicious activity. Most of the time witnesses do not come forward to make a report because they try to avoid conflict with others. The rules of the organization should provide complete protection from retribution and anonymity to witnesses, if required.

### Reporting Methods to Consider

The reporting form should be easy and quick to complete (see Appendix G for a sample incident report form). Taking an hour or two to fill in a tedious form may discourage workers and students from using the reporting system. As part of their basic and continuing safety training, all workers and students need instruction on when and how to fill out the form. The designated safety committee for the organization should establish procedures to receive reports and take appropriate action in a timely fashion. Anonymous filing of incident reports should be considered. There should be a secure place, a designated third party, or a Web site for filing reports of incidents, so that people who are reporting questionable safety actions are assured of confidentiality.

The purpose of filing incident forms is not to attribute blame but to make it possible for the CSO, scientific leaders, and administrators to address basic safety problems and to add to or modify the rules for laboratory safety and security.

### Investigations

An investigation should be used to establish the facts of an incident, determine the cause of a problem, and recommend improvements. All incidents should be investigated, but the depth of each investigation is determined by the seriousness of the incident, according to a process established by the safety committee. For example, a minor incident may require only a call or short interview with an individual or group. The findings of all investigations should be in writing.

## APPROACHES TO FOSTERING COMPLIANCE

Changing behaviors and fostering a culture of compliance are challenging. Local social and cultural barriers may inhibit a laboratory manager, laboratory personnel, students, and others from complying with the best safety and security practices. This section discusses approaches that can be implemented to change noncompliant behaviors and improve laboratory safety and security. In addition to what is written here, a laboratory manager will need to utilize educational tools to foster compliance. For example, case studies can be developed and used to train laboratory personnel. An example case study "Ensuring the Use of Safety Measures in the Laboratory" is provided in Appendix G.

### Setting Organizational Safety Rules, Policies, and Implementation Strategy

Good compliance requires clear rules, policies, and processes that have been agreed on by organizational leaders, safety and security officers, and laboratory managers. It is also critical for compliance and administration that key stakeholders in the organization also agree to a clear, direct strategy for implementing rules. Rules need to be approved by the highest forums, such as a board of governors or trustees, if they are to have legitimacy and be legally binding. Rules need to be printed and circulated as official organizational documents from the office of the chancellor or president.

### Dealing with Limited Financial Resources

Maintaining and improving a system require sustained financial support. However, increasing safety does not have to be expensive. Strong leadership can lead to changes in personal behavior that can result in improved chemical safety and security. Changing personal behavior can be an effective and inexpensive way to improve chemical safety and security.

### Addressing Climate Control

In some developing countries, heat and humidity are excessive during most of the year and mechanical ventilation and air conditioning are unavailable. Appropriate actions should be taken to keep people comfortable while they comply with safety practices and rules. For example, one university in the Philippines made it possible to work in humid conditions by purchasing antifog chemical splash goggles for laboratory workers.

### Providing Training and Education

People require training to become aware of potential hazards. No one should be allowed to work in chemical laboratories without adequate training in laboratory standard operating procedures. Laboratory personnel should be comfortable asking safety and security officers for expert advice on what to do, before they proceed with risky actions. Safety and security officers should have updated and adequate knowledge to guide others. In developing countries, those officers can be sent to civil defense organizations or other public agencies for training. Scientific leaders, safety and security officers, and others in authority need to be careful when writing directions and instructions they distribute. Material that is distributed should be checked for accuracy and thoroughness. Sloppy, offhand, or ill-informed instructions can be harmful.

### Encouraging Rest and Well-being

Working while physically or mentally tired is one of the most common causes of laboratory accidents, near-miss incidents, and lapses of security. Workers and students need to look out for each other and encourage ill or exhausted coworkers to leave the laboratory and get rest or sleep so that they will be able to meet the stress and effort of work. The organization should support workers and students in participating in interesting, extracurricular activities on a regular basis to reduce mental stress and achieve a more balanced life. Happy, rested workers make an organization productive and safe.

### Enforcing Consequences of Risky Behavior

Rules for safe behavior and penalties for their violation should be widely publicized, in advance, to make everyone aware of them. If people know that negligent or deliberately risky behavior in laboratories or breaches in security will have no consequences, they will have little incentive to change their habits. Consequences of safety or security violations could include publicity of the violations, restrictions on use of laboratory facilities and equipment, monetary fines, withdrawal of financial support, or job termination. Consequences should be proportional to the severity of the violations. To promote compliance with rules, leaders also need to reward people who have consistently taken safe actions and behaved responsibly. A reward might be monetary or simply favorable recognition.

### Relieving Time Pressures and Avoiding Shortcuts

Trying to do laboratory or experimental processes too fast can lead to mistakes and accidents or incidents. Shortcuts in standard operating procedures can compromise safety. Supervisors and laboratory leaders need to be mindful of the time required to complete assigned work and of the risks and consequences that can ensue if they reduce time without adding workers. In designing experiments, supervisors should consult with workers to validate the proper allocation of time required for every step of an experiment. Adequate time is needed to do things the right way. Additional education and training may be required to give people incentives to avoid dangerous shortcuts. Every person in a laboratory should learn about the consequences of shortcuts and be made aware of the penalties for taking them. Coworkers should learn to encourage one another to work safely.

### Taking Special Safety Precautions for Women

Women require additional safety measures to protect their reproductive health. For example, certain chemicals are reproductive toxins that women should not handle. Organizational leaders should ensure that female laboratory personnel are provided with the appropriate guidelines, training, and equipment needed for their safety and security.

In addition, cultural or religious traditions could keep men from giving women physical assistance that they need in emergencies. In case of such situations, laboratory safety offices and security offices should hire women.

### Accommodating Social, Ethnic, and Religious Differences

Discrimination against groups and against persons of low social status happens globally. Institutions should have clearly defined policies with regard to fair treatment of all workers. Lower-status workers in particular are often involved in cleaning and other potentially hazardous jobs, but have little or no education in chemistry. They should be provided with adequate PPE and training to avoid harm to their health in the line of duty. Scientific leaders and administrators need to become role models for equitable, objective, and humane treatment of all workers and students. In some cases, leaders may be legally obligated to take such measures. Large personal fines or even prison sentences may be implemented if leaders do not provide a safe and secure working environment for students and staff.

### Accommodating Propriety in Dress and Behavior

All laboratory members should be educated about and kept aware of the need to wear proper clothing and protective equipment. They should have ready access to proper clothing for the laboratory such as lab coats and gloves even if they prefer to wear traditional clothing outside. Feelings and traditional standards of propriety may discourage persons, particularly women, who have been splashed with caustic chemicals or other hazardous materials in the laboratory from immediately removing contaminated clothing to reduce chemical burns and from going to and using emergency safety showers properly. It may be necessary for educational institutions to provide laboratory sessions for female students that are separated in time and possibly location from those for male students, or to specially design personal protective clothing and equipment that can accommodate fitting under or over traditional attire.

### Confronting Coworkers or Superiors

Laboratory workers may witness safety or security breaches but be fearful of or apprehensive about confronting coworkers and authorities. These are normal feelings and reactions that should be countered by providing anonymity for informants, if possible, by protecting informants and by preventing reprisals.

Proper handling of such a situation depends heavily on having clear, agreed rules and an objective, fair, well-publicized, and understood strategy for investigating incidents and administering the consequences of breaking or disregarding the rules. The messenger should not be blamed but rather thanked for rendering a valuable service. Please refer to earlier section on protection for those who report incidents.

### Looking Out for Coworkers

A person's sense of survival or concern for the well-being of others can be complicated. Specific rules or strong guidance may be needed for training workers on when and how to help others and one's self in emergencies, and even more importantly, on when to cooperate with others to prevent accidents and emergencies. All laboratory workers and students should also receive adequate education on the importance of both wearing PPE and training in its proper use; these are critical for compliance with laboratory safety rules.

# A

# Statement of Task

An ad hoc committee of the National Research Council will undertake the following tasks:

1). It will produce materials (as noted below) providing guidance on a baseline of practices required to promote safety and security in their handling and use of toxic industrial chemicals (TICs) and other hazardous chemicals on the laboratory scale in the developing world. It will:

- Consider current safety and security practices in these countries based on information from the Department of State Chemical Security Engagement Program, and other organizations such as the International Union of Pure and Applied Chemistry, and the American Chemical Society Committee on International Affairs. Use this information to determine practices most needed and/or most readily employed in developing economies.
- Based on information found in *Prudent Practices in the Laboratory: Safe Handling and Disposal of Chemical* (NAP, 1995), produce materials (booklets, CDs, or other media) that outline basic steps, feasible in developing economies, to improve chemical management best practices, including enhanced safety and security in the use, storage, and disposal of hazardous chemicals. This should include consideration of training and other "culture of safety" issues.

2). It will examine the dual risks—in particular, the risk posed by theft and diversion of relatively small amounts of chemicals from laboratory settings—posed by TICs and other hazardous chemicals in developing countries, particularly in regions where terrorism is on the rise. In its final report, it will provide guidance on a baseline of practices required to promote good chemical management practices to ensure safety and security in their handling and use in laboratories in developing world. Specifically, this study will:

- Examine current patterns of use and distribution of chemicals in the developing world, especially countries where terrorism is of particular concern, describing, in general terms, the types and levels of laboratory activities and their geographic distribution.
- Examine current safety and security practices and attitudes in laboratories based on information from the Department of State Chemical Security Engagement Program, international organizations such as the International Union of Pure and Applied Chemistry, organizations such as the American Chemical Society Division of International Affairs and the American Chemistry Council, and members of the chemical community operating abroad, including both academic and industrial practitioners. Identify practices that provide the greatest opportunity to improve safety and security and/or practices most amenable to readily-employed mitigation techniques.
- Examine on-going efforts to engage chemical professionals (scientists, technicians and engineers) from the developing world with the international R&D community in order to improve best practices in chemical safety and security.
- Recommend basic steps, feasible in developing economies, to provide enhanced safety and security in the use, storage, and disposal of hazardous chemicals. This should include consideration of training opportunities to engage chemical professionals from developing countries in activities to improve best practices in chemical safety and security and development of long-term relationships that could foster improved security.

# B

# Meeting Agenda

**MEETING 1**
**March 2, 2009**
**Washington, D.C.**

| | |
|---|---|
| 10:00 a.m. | **Welcome and Introductions**<br>*Ned Heindel*, Chair |
| | CHEMICAL SECURITY ENGAGEMENT PROGRAM |
| 10:15 a.m. | **Program Overview and Discussion of Study Charge**<br>*Marie Ricciardone*, U.S. Department of State |
| 11:15 a.m. | **CSP Chemical Safety and Security Training**<br>*Nancy Jackson*, Sandia National Laboratories |
| 12:15 p.m. | Lunch |
| 1:15 p.m. | **CSP Training Materials**<br>*Nancy Jackson*, Sandia National Laboratories |
| 2:15 p.m. | **Discussion of Proposed Educational Materials with Guests** |
| 3:15 p.m. | **Break** |
| 3:30 p.m. | Closed Session |

**March 3, 2009**

CLOSED SESSION

**MEETING 2**
**April 27-28, 2009**
**Washington, D.C.**

CLOSED SESSION

**MEETING 3**
**June 4, 2009**
**Washington, D.C.**

| | |
|---|---|
| 8:15 a.m. | **Welcome and Introductions**<br>*Ned Heindel*, Chair |
| 8:30 a.m. | (video conference) Dr. Mohammad El-Khateeb, Chairman, Department of Chemistry, Jordan University of Science and Technology |
| 9:30 a.m. | Dr. Temechegn Engida, President, Federation of African Societies of Chemistry, Addis Ababa, Ethiopia |
| 10:30 a.m. | Break |
| 10:45 a.m. | Dr. Khalid Temsamani, National Coordinator, Materials Science; Professor of Electro-Analytical Chemistry in the Faculty of Sciences of Tetouan, University Abdelmalek Essaadi, Morocco |
| 11:45 a.m. | Dr. Supawan Tantayanon, President, Thai Chemical Society; Associate Dean for Academic Affairs, Faculty of Science, Chulalongkorn University. |
| 12:15 p.m. | Lunch |

| | |
|---|---|
| 2:15 p.m. | Dr. Richard W. Niemeier, Senior Scientist and Toxicologist, Associate Director of Science, Education and Information Division, U.S. National Institute for Occupational Safety and Health |
| 3:30 p.m. | Dr. Alastair W.M. Hay, Professor of Environmental Toxicology<br>University of Leeds, United Kingdom; Chair, IUPAC/OPCW Multiple Uses of Chemistry Workgroup |
| 4:30 p.m. | Mr. Francisco Gomez, ACS Office of International Activities |
| 5:00 p.m. | **Closing Discussion** |
| 5:30 – 7:30 p.m. | **Dinner for Committee and Guests** |

**June 5, 2009**

CLOSED SESSION

MEETING 4
**July 16-17, 2009**
Woods Hole, Massachusetts

CLOSED SESSION

# C

# Committee and Guest Speaker Information

## COMMITTEE MEMBERS

**Ned D. Heindel** *(Chair)* is H. S. Bunn Chair and Professor of Chemistry at Lehigh University. He joined Lehigh University in 1966. Dr. Heindel's research has focused mainly on the preparation of radiopharmaceuticals and synthesis of useful therapeutic drug candidates. He is working on countermeasures for sulfur mustard vesicant. Dr. Heindel has 11 patents, four of which have been licensed. In 1994 he served as the president of the American Chemical Society. Dr. Heindel earned his B.S. at Lebanon Valley College in 1959 (chemistry and mathematics) and his Ph.D. at the University of Delaware in 1963 (organic chemistry), and he held a postdoctoral fellowship at Princeton University in 1964 (medicinal chemistry).

**Charles Barton** is an independent consultant. His expertise includes cause–effect and dose–response relationships. He was until recently a senior scientist at XOMA (US) LLC, in Berkeley, California, where he oversaw preclinical studies to determine the toxicity and safety of therapeutic monoclonal antibodies. Before joining XOMA, Dr. Barton was the state toxicologist for Iowa. He has served on the faculties of Iowa State University and Des Moines University Osteopathic Medical Center. Dr. Barton received his Ph.D. in toxicology from the University of Louisiana. In addition to being a board-certified toxicologist, he is certified in conducting public health assessments, health education activities, and risk assessments; emergency response to terrorism and emergency response incident command; and hazardous waste operations and emergency response.

**Janet S. Baum** has focused her professional career on complex R&D facilities for medical, biotechnology, pharmaceutical, and academic clients. Her specialized expertise includes planning for animal facilities, molecular- and cellular-biology laboratories, and biosafety laboratories (to level 4). Ms. Baum works with researchers, scientists, facilities staff, health and safety personnel, and administrators to understand "big picture" objectives, create consensus, and develop project requirements through understanding of scientific processes and functions. Ms. Baum teaches at the Harvard University School of Public Health and Washington University in St. Louis. She is widely published on laboratory health and safety guidelines. She is the author or coauthor of 15 books and numerous articles.

**Apurba Bhattacharya** is associate professor of chemistry at Texas A&M University-Kingsville. Recently he held the position of senior vice president and head of global R&D in Dr. Reddy's Laboratories Ltd. in Hyderabad, India. Dr. Bhattacharya joined the faculty of Texas A&M after many years in the pharmaceutical industry, where he started his career with Merck & Co. as a senior research chemist in process R&D. Later he joined Hoechst, where he rose to lead chemist of the Innovator Group. He then held the position of group leader in central process research at Bristol Myers Squibb until joining the faculty of Texas A&M in 1999. During his time in industry, Dr. Bhattacharya developed the synthesis of Propecia™ (hair growth drug) and Proscar™ (for benign prostatic hypertrophy). He worked on chiral Robinson annulation, and the synthesis of *S*-ibuprofen, D-*p*-hydroxyphenyl glycine, cromolyn sodium, alkyl indanones, quinazolinones, amphoteric copolymer, and MRI imaging agents. Dr. Bhattacharya's current research interests include designing environmentally benign, waste-free chemistry with emphasis on current industrial synthesis and processes using homogeneous and heterogeneous catalyst systems and asymmetric synthesis, and chiral phase-transfer catalysis. Dr. Bhattacharya received his Ph.D. in organic chemistry from the University of Texas at Austin in 1982; his M.S. in chemistry from the Indian Institute of Technology, Kanpur, India, in 1976; and his B.S. in chemistry from Calcutta University, India, in 1974. Dr. Bhattacharya's contributions have extended to numerous newspaper articles, 90 refereed publications, and 26 patents.

**Charles P. Casey** (member, National Academy of Sciences) is Homer B. Adkins Professor Emeritus of Chemistry at the University of Wisconsin-Madison. His research lies at the interface of organometallic chemistry and homogeneous catalysis; his group studies the mechanisms of homogeneously catalyzed reactions. He received his B.S. from St. Louis University and his Ph.D. from the Massachusetts Institute of Technology.

**Mark C. Cesa** is a senior research associate for INEOS USA LLC. He received a Ph.D. (1979) and an M.S. (1977) in organic chemistry from the University of Wisconsin-Madison and an A.B. in chemistry from Princeton University (1974). Dr. Cesa is a past chair of the U.S. National Committee for the International Union of Pure and Applied Chemistry (IUPAC). He is chair of the IUPAC Committee on Chemistry and Industry, which conducts the Safety Training Program sponsored by IUPAC, the U.N. Educational, Scientific and Cultural Organization (UNESCO) and the U.N. International Development Organization (UNIDO). The program allows safety experts from developing countries to learn more about safety and environmental protective measures by visiting and working in plants of IUPAC company associates in the industrialized world. IUPAC, UNESCO, and UNIDO established and have maintained the Safety Training Program to promote interactions between developed countries and the developing world to disseminate state-of-the-art knowledge on safety and environmental protection in chemical production.

**M. Iqbal Choudhary** is the director of the International Center for Chemical and Biological Sciences and the Dr. Panjwani Center for Molecular Medicine and Drug Research, University of Karachi, Pakistan. Dr. Choudhary obtained his Ph.D. from the H.E.J. Research Institute of Chemistry in 1987 and his M.Sc. in 1983 from the University of Karachi, Pakistan, also in organic chemistry. He received his B.Sc. from the University of Karachi in chemistry, biochemistry, and botany. Dr. Choudhary is involved in academic projects, including a survey of medicinal plants in Pakistan; environmental monitoring; and capacity building in science and technology in Pakistan. Dr. Choudhary is a member of the Royal Society of Chemistry, London; the American Chemical Society; the International Union of Pure and Applied Chemistry; the American Society of Pharmacology; the New York Academy of Sciences; and the Federation of Asian Chemical Societies. He was awarded the Tamgha-E-Imtiaz in 1998, the Sitara-E-Imtiaz in 2001, and the Hilal-e-Imtiaz in 2006, all by the president of Pakistan. He was also awarded the Abdussalam Prize in Chemistry in 1990 and the Young Chemist Award of TWAS, the Academy of Sciences for the Developing World in 1994. Dr. Choudhary was elected a fellow of the Islamic Academy of Sciences in 2002.

**Robert H. Hill** is a program manager with Battelle in Atlanta, GA, where he manages several contracts with the Centers for Disease Control and Prevention (CDC). Formerly Dr. Hill worked in or managed laboratories for 30 years, and he has more than 30 years of experience in occupational and environmental health at CDC. Dr. Hill has served as member-at-large, chair of the American Chemical Society (ACS) Division of Chemical Health and

Safety and is an active member of the ACS Executive Committee. He also serves as liaison to the ACS Committee on Chemical Safety. He has served as member of the Board of Editors of the *Journal of Chemical Health and Safety* since 2000. He served as the ACS representative member and later president of the National Registry of Certified Chemists. Dr. Hill received his Ph.D. in chemistry from the Georgia Institute of Technology and his B.S. in chemistry from Georgia State University. He received the Howard Fawcett Award for outstanding achievements in chemical health and safety.

**Robin Izzo** is the associate director for laboratory safety in the Princeton University Office of Environmental Health and Safety. She has more than 20 years of experience in laboratory safety, having held positions at the University of Vermont and Harvard University before her 16-year tenure at Princeton. Ms. Izzo was instrumental in working with the U.S. Environmental Protection Agency (EPA) in developing proposed rule making to make compliance with chemical waste regulations more relevant to colleges and universities. Ms. Izzo is the chair of the coordinating committee for the EPA College and University Sector Strategy, coordinating the efforts of six national and international organizations to develop a framework for environmental compliance and sustainability programs at colleges and universities. She is a member of the Board of Directors of the Campus Safety Health and Environmental Management Association. Ms. Izzo holds a B.S. in mathematics from the University of Vermont and an M.S. in environmental sciences from the New Jersey Institute of Technology.

**Patrick J. Y. Lim** is professor and chair of the Department of Chemistry in the University of San Carlos in Cebu City, Philippines. Through an Australian Development Cooperation scholarship, he completed his Ph.D. in chemistry (2000) at the University of Melbourne under the supervision of Charles G. Young. His doctoral work investigated novel reactions of metal and sulfur compounds with activated alkynes and produced five publications in such journals as *Inorganic Chemistry* and *Organometallics*. On returning to the Philippines, he rose in the ranks of the department, becoming chair in 2004. He was recently appointed editor of *The Philippine Scientist*, a multidisciplinary ISI journal published by the university. He serves as an accreditor of the Philippines Accrediting Association of Schools, Colleges and Universities and sits on the Philippine Commission on Higher Education's Technical Committee for Chemistry.

**Russell W. Phifer** is the principal of WC Environmental, LLC. He has over 25 years of experience in environmental health and safety (EH&S). His background includes management of health and safety at Superfund sites, training of chemists in safety, and consulting on environmental health and safety issues for laboratory and industrial facilities. Mr. Phifer has received professional certification from a variety of professional organizations, and is an Occupational Safety and Health Administration authorized trainer. He has served in numerous capacities for the American Chemical Society (ACS) and is immediate past chair of the ACS Committee on Chemical Safety, and chair of the ACS Division of Chemical Health and Safety. Mr. Phifer has served actively with the ACS Laboratory Chemical and Waste Management Task Force since 1981, including six years as chair. He is a member of the Board of Editors of the *Journal of Chemical Health and Safety*. He currently serves as a member of the National Research Council's Committee on Prudent Practices for the Handling and Disposal of Chemicals.

**Mildred Solomon** is vice president of the Education Development Center Inc. (EDC), an international nonprofit R&D organization of more than 1,200 professional staff, and associate clinical professor of medical ethics and anaesthesia at Harvard Medical School. Dr. Solomon directs the EDC Center for Applied Ethics, an interdisciplinary group of social scientists engaged in a variety of studies focusing on values questions in medicine and health care and on health system quality improvement. At Harvard, she directs the medical school's Fellowship in Medical Ethics, a program aimed at building the bioethics capacity of Harvard-affiliated teaching hospitals. An expert in ethics education and behavioral change, Dr. Solomon has more than 30 years experience in researching, designing, and evaluating education and quality improvement programs for health professionals, health care organizations, and the public, particularly in medical uncertainty, in which values questions pose policy and practice challenges. She received her B.A. from Smith College and her doctorate from Harvard University.

**James M. Solyst,** a principal consultant with ENVIRON, has more than 25 years of experience in advising businesses and policy leaders on the application of science in decision making and communicating science to key audiences, including regulatory and legislative bodies. Mr. Solyst is experienced in product stewardship, global chemical management, emergency response, and corporate responsibility. He has assisted U.S. governors with initiatives and incidents through the National Governors Association and chemical companies responding to emerging science through the American Chemistry Council. He has also worked on international initiatives, including REACH, the U.N. Environmental Programme's Strategic Approach to International Chemicals Management, and the harmonization of global product stewardship programs. Mr. Solyst is a member of the American Chemical Society Committee on Environmental Improvement, and he is an external affiliate of the Johns Hopkins School of Public Health Risk Sciences and Public Policy Institute. He received his M.S. in city and regional planning from Ohio State University and his B.A. from the University of Maryland.

**Usha Wright** is executive vice president and co-general counsel for O'Brien & Gere, an environmental engineering and consulting firm in New York. She has extensive international industry experience in chemical safety. In 2008 she retired as senior vice president for global workforce strategy at ITT Corporation, a position she had held since 2005. From 1993 to 2005 Ms. Wright was vice president and associate general counsel for ITT, with responsibility for environment, safety, and health (ES&H). Before joining ITT, she was executive director of environmental health and safety at Ciba Geigy Pharmaceuticals from 1977 to 1993. Ms. Wright has a B.S. in chemistry from Rutgers University, an M.S. from the University of North Carolina, and a J.D. from Rutgers University. She is a Certified Industrial Hygienist and a Certified Safety Professional. She is on the board of the Environmental Law Institute, where she is involved in conducting training in ES&H compliance in various academic institutions in India. She is also on the board of SHARE (*shareafrica.org*), a nongovernmental organization working in western Kenya.

## GUEST SPEAKERS

**Mohammad El-Khateeb** is the chairman of the Chemistry Department of the Jordan University of Science and Technology in Irbid. He has held this position since 2007, and he joined the department in 1996. Dr. El-Khateeb served as the vice dean of the Faculty of Science and Arts at the university from 2002 to 2004. He received a B.S. (1988) and an M.S. (1990) in chemistry from Yarmouk University in Irbid, Jordan, and a Ph.D. in inorganic and organometallic chemistry from McGill University in Montreal, Canada, in 1996. He has received numerous awards, including the 2008 Abdul Hameed Shoman Award for Young Arab Researchers in chemistry and the 2004–2005 Alexander von Humboldt Scholarship at the Friedrich Schiller University of Jena in Germany. He serves as the treasurer of the Jordanian Chemical Society.

**Temechegn Engida** received a B.S. in chemistry (1988) and an M.A. in chemical education (1993) from Addis Ababa University. He received a Ph.D. in chemical education from the University of Muenster in Germany in 2000, specializing in structural chemistry education (with emphasis on the structures of solids). Since earning his B.S., Dr. Temechegn has been lecturing at Addis Ababa University and advising students working in chemical education at the undergraduate, master's, and Ph.D. levels. He has been researching and publishing articles on chemical education. From August 2004 to February 2007, he served as the vice president of the Chemical Society of Ethiopia. During that time, he played a key role in initiating and founding the Federation of African Societies of Chemistry (FASC) in February 2006. He is now leading FASC as its founding president. He also works for the UNESCO-International Institute for Capacity Building in Africa, based in Addis Ababa.

**Francisco Gomez** works in the American Chemical Society (ACS) Office of International Activities. He is responsible for developing and implementing international alliance and partnership opportunities for ACS and fostering existing ones. Before joining ACS, Mr. Gomez worked as a consultant at G&G Consulting, advising Latin American clients on strategic planning and the development of programs aimed at maximizing organizational effectiveness. Before joining G&G, Mr. Gomez served as district manager for Healthcare Services Group Inc., where he was responsible for all aspects of operations, including financial control, business development, operations, client relations, regulatory compliance, and human capital. Mr. Gomez holds a B.S. in business administration from Marshall University with a concentration in management and economics and an M.B.A. from the Kogod School of Business of American University with a concentration in international business. A native of Colombia, he is fluent in Spanish and has considerable knowledge of developing countries, having studied and worked in Latin America.

**Alastair Hay** is a professor of environmental toxicology at the University of Leeds. He has a B.S. in chemistry (1969) and a Ph.D. in biochemistry (1973) from London University. Most of his research is on the effects of chemicals on health, and he has published many papers and articles in the scientific and medical press. Since 1989, Dr. Hay has been a member of a number of U.K. government committees that have made recommendations on the regulation of chemicals and occupational exposure standards. Dr. Hay has also worked on issues related to chemical and biologic warfare for some 30 years. Much of his work has dealt with the need for workable and international treaties that prevent the use of warfare with chemical or biologic agents. Dr. Hay recently chaired a small international working group under the auspices of the International Union of Pure and Applied Chemistry and the Organization for the Prohibition of Chemical Weapons that prepared educational material for chemists on the multiple uses of chemicals, chemical warfare, and codes of conduct. Dr. Hay is a partner in a project coordinated by Environment Canada to produce protocols for cleaning surfaces after chemical- or biologic-agent contamination.

**Richard W. Niemeier** is a toxicologist who has worked at the National Institute for Occupational Safety and Health (NIOSH) for 33 years. His experience includes work with the International Programme on Chemical Safety of the World Health Organization (WHO) over the last 22 years to produce more than 1,600 international chemical safety cards. He was instrumental in bringing the message of control banding from the U.K. Health and Safety Executive to NIOSH and the United States. Dr. Niemeier serves as a deputy manager for a segment of the WHO Global Collaborating Centre Network. The network has worked with South America and Mozambique in taking the control banding concept to developing countries by using the WHO toolkit. In addition, he was a member and chair of a subcommittee of the Federal Bureau of Investigation's Scientific Working Group on the Forensic Analysis of Chemical Terrorism. For the last 11 years, Dr. Niemeier has been a member of the EPA Federal Advisory Committee on Acute Exposure Guideline Levels, and he was recently appointed to the EPA Integrated Risk Information System Federal Standing Science Committee.

**Supawan Tantayanon** is the director of the Technopreneurship and Innovation Management and associate professor of chemistry in the Faculty of Science at Chulalongkorn University, Thailand. She first joined the chemistry department as an instructor in 1975. She is also an affiliate associate professor in the Department of Chemistry and Biochemistry at Worcester Polytechnic Institute. Dr. Tantayanon's interests and expertise include research in organic and polymer synthesis, green chemistry, alternative energy, and educational topics in chemical safety, green chemistry, and small-scale chemistry. She received a B.S. in chemistry (1973) from Chulalongkorn University and an M.S. in organic chemistry (1975) from Mahidol University, Thailand. She received a Ph.D. in organic chemistry from Worcester Polytechnic Institute in 1982 and a diploma in polymer science from Ferrara University, Italy, in 1993. Dr. Tantayanon has held numerous national and international positions, including being the president of the Pacific Polymer Federation in 2002–2003, president of the Polymer Society of Thailand in 1997–2003, director of the Green Chemistry Institute in Thailand since 2002, and president of the Chemical Society of Thailand since 2007. Dr. Tantayanon is the 2009 president-elect of the Federation of Asian Chemical Societies.

**Khalid R. Temsamani** is the national coordinator for materials science and professor of electroanalytic chemistry at the Faculty of Sciences of Tetouan, University of Abdelmalek Essaâdi, Morocco. He received his Ph.D. in chemistry in 1988 from the Université Libre de Bruxelles, Belgium. He is certified in immunology and cancerology (1987). In 2008 he became director of the Materials and Interfacial Systems Laboratory. Dr. Temsamani is the representative of Morocco to the U.S. National Science Foundation and he recently joined Morocco's National Biosecurity Council and the MENA Region Core Group for Biosafety and Biosecurity. He is an adviser on biosafety, biosecurity, and science ethics and served as a consultant to the U.S. National Academies to conduct a study of Morocco's capabilities in biosafety and biosecurity. He has supervised more than 15 graduate research works and is author of 32 research papers and 74 international and national communications.

# D

# Examples of Chemicals of Concern

The following Tables (D-1 through D-5) are examples of the types of chemicals that a laboratory would include in an inventory of Chemicals of Concern (COCs).

**TABLE D-1** Chemical Weapons and Chemical Weapons Precursors[a]:

| Chemical of Concern | Synonym | CAS Registry Number[b] |
|---|---|---|
| 1,4-Bis(2-chloroethylthio)-*n*-butane | | 142868-93-7 |
| Bis(2-chloroethylthio)methane | | 63869-13-6 |
| Bis(2-chloroethylthiomethyl)ether | | 63918-90-1 |
| 1,5-Bis(2-chloroethylthio)-*n*-pentane | | 142868-94-8 |
| 1,3-Bis(2-chloroethylthio)-*n*-propane | | 63905-10-2 |
| 2-Chloroethylchloro-methylsulfide | | 2625-76-5 |
| Chlorosarin | O-Isopropyl methylphosphonochloridate | 1445-76-7 |
| Chlorosoman | O-Pinacolyl methylphosphonochloridate | 7040-57-5 |

*continued*

**TABLE D-1** Continued

| Chemical of Concern | Synonym | CAS Registry Number[b] |
|---|---|---|
| DF | Methyl phosphonyl difluoride | 676-99-3 |
| Ethyl phosphonyl difluoride | | 753-98-0 |
| HN1 (nitrogen mustard-1) | Bis(2-chloroethyl)ethylamine | 538-07-8 |
| HN2 (nitrogen mustard-2) | Bis(2-chloroethyl)methylamine | 51-75-2 |
| HN3 (nitrogen mustard-3) | Tris(2-chloroethyl)amine | 555-77-1 |
| Isopropylphosphonyl difluoride | | 677-42-9 |
| Lewisite 1 | 2-Chlorovinyldichloroarsine | 541-25-3 |
| Lewisite 2 | Bis(2-chlorovinyl)chloroarsine | 40334-69-8 |
| Lewisite 3 | Tris(2-chlorovinyl)arsine | 40334-70-1 |
| Sulfur mustard (mustard gas (H)) | Bis(2-chloroethyl)sulfide | 505-60-2 |
| O-Mustard (T) | Bis(2-chloroethylthioethyl)ether | 63918-89-8 |
| Propylphosphonyl difluoride | | 690-14-2 |
| QL | O-Ethyl-O-2-diisopropylaminoethyl methylphosphonite | 57856-11-8 |
| Sarin | O-Isopropyl methylphosphonofluoridate | 107-44-8 |
| Sesquimustard | 1,2-Bis(2-chloroethylthio)ethane | 3563-36-8 |
| Soman | O-Pinacolyl methylphosphonofluoridate | 96-64-0 |
| Tabun | O-Ethyl-N,N-dimethylphosphoramido-cyanidate | 77-81-6 |
| VX | O-Ethyl-S-2-diisopropylaminoethyl methyl phosphonothiolate | 50782-69-9 |

NOTE: Toxic chemicals with few or no legitimate uses, developed or used primarily for military purposes.

[a]U.S. Chemical Weapons Convention Schedule 1; see *http://www.cwc.gov/* (accessed October 28, 2009).

[b]See Chemical Abstract Service web site *www.cas.org* (accessed October 28, 2009).

SOURCE: U.S. Department of Homeland Security list of Chemicals of Interest (6 CFR Part 27 Appendix to Chemical Facility Anti-Terrorism Standards; Final Rule; November 20, 2007).

**TABLE D-2** Explosives and Improvised Explosive Device Precursors

| Chemical of Concern | Synonym | CAS Registry Number |
|---|---|---|
| Aluminum (powder) | | 7429-90-5 |
| Ammonium nitrate | | 6484-52-2 |
| Ammonium perchlorate | | 7790-98-9 |
| Ammonium picrate | | 131-74-8 |
| Barium azide | | 18810-58-7 |
| Diazodinitrophenol | | 87-31-0 |
| Diethyleneglycol dinitrate | | 693-21-0 |
| Dingu | Dinitroglycoluril | 55510-04-8 |
| Dinitrophenol | | 25550-58-7 |
| Dinitroresorcinol | | 519-44-8 |
| Dipicryl sulfide | | 2217-06-3 |
| Dipicrylamine [or] Hexyl | Hexanitrodiphenylamine | 131-73-7 |
| Guanyl nitrosaminoguanylidene hydrazine | | |
| Hexanitrostilbene | | 20062-22-0 |
| Hexolite | Hexotol | 121-82-4 |
| HMX | Cyclotetramethylene-tetranitramine | 2691-41-0 |
| Hydrogen peroxide (concentration of at least 35%) | | 7722-84-1 |
| Lead azide | | 13424-46-9 |
| Lead styphnate | Lead trinitroresorcinate | 15245-44-0 |
| Magnesium (powder) | | 7439-95-4 |
| Mercury fulminate | | 628-86-4 |
| Nitrobenzene | | 98-95-3 |
| 5-Nitrobenzotriazol | | 2338-12-7 |
| Nitrocellulose (not filters) | | 9004-70-0 |
| Nitroglycerine | | 55-63-0 |
| Nitromannite | Mannitol hexanitrate, wetted | 15825-70-4 |
| Nitromethane | | 75-52-5 |
| Nitrostarch | | 9056-38-6 |
| Nitrotriazolone | | 932-64-9 |
| Octolite | | 57607-37-1 |
| Octonal | | 78413-87-3 |
| Pentolite | | 8066-33-9 |
| PETN | Pentaerythritol tetranitrate | 78-11-5 |

*continued*

**TABLE D-2** Continued

| Chemical of Concern | Synonym | CAS Registry Number |
|---|---|---|
| Phosphorus | | 7723-14-0 |
| Potassium chlorate | | 3811-04-9 |
| Potassium nitrate | | 7757-79-1 |
| Potassium perchlorate | | 7778-74-7 |
| Potassium permanganate | | 7722-64-7 |
| RDX | Cyclotrimethylenetrinitramine | 121-82-4 |
| RDX and HMX mixtures | | 121-82-4 |
| Sodium azide | | 26628-22-8 |
| Sodium chlorate | | 7775-09-9 |
| Sodium nitrate | | 7631-99-4 |
| Tetranitroaniline | | 53014-37-2 |
| Tetrazene | Guanyl nitrosaminoguanyltetrazene | 109-27-3 |
| 1*H*-Tetrazole | | 288-94-8 |
| TNT | Trinitrotoluene | 118-96-7 |
| Torpex | Hexotonal | 67713-16-0 |
| Trinitroaniline | | 26952-42-1 |
| Trinitroanisole | | 606-35-9 |
| Trinitrobenzene | | 99-35-4 |
| Trinitrobenzenesulfonic acid | | 2508-19-2 |
| Trinitrobenzoic acid | | 129-66-8 |
| Trinitrochlorobenzene | | 88-88-0 |
| Trinitrofluorenone | | 129-79-3 |
| Trinitro-*m*-cresol | | 602-99-3 |
| Trinitronaphthalene | | 55810-17-8 |
| Trinitrophenetole | | 4732-14-3 |
| Trinitrophenol | Picric acid | 88-89-1 |
| Trinitroresorcinol | | 82-71-3 |
| Tritonal | | 54413-15-9 |

SOURCE: U.S. Department of Homeland Security list of Chemicals of Interest (6 CFR Part 27 Appendix to Chemical Facility Anti-Terrorism Standards; Final Rule; November 20, 2007).

TABLE D-3 Weapons of Mass Effect Agents and Precursors

| Chemical of Concern | Synonym | CAS Registry Number |
|---|---|---|
| Arsine | | 7784-42-1 |
| Boron tribromide | | 10294-33-4 |
| Boron trichloride | Borane, trichloro | 10294-34-5 |
| Boron trifluoride | Borane, trifluoro | 7637-07-2 |
| Bromine chloride | | 13863-41-7 |
| Bromine trifluoride | | 7787-71-5 |
| Dinitrophenol | | 25550-58-7 |
| Dinitroresorcinol | | 519-44-8 |
| Carbonyl fluoride | | 353-50-4 |
| Chlorine pentafluoride | | 13637-63-3 |
| Chlorine trifluoride | | 7790-91-2 |
| Cyanogen | Ethanedinitrile | 460-19-5 |
| Cyanogen chloride | | 506-77-4 |
| Diborane | | 19287-45-7 |
| Dichlorosilane | Silane, dichloro- | 4109-96-0 |
| Dinitrogen tetroxide | | 10544-72-6 |
| Fluorine | | 7782-41-4 |
| Germane | | 7782-65-2 |
| Germanium tetrafluoride | | 7783-58-6 |
| Hexafluoroacetone | | 684-16-2 |
| Hydrogen bromide (anhydrous) | | 10035-10-6 |
| Hydrogen chloride (anhydrous) | | 7647-01-0 |
| Hydrogen cyanide | Hydrocyanic acid | 74-90-8 |
| Hydrogen fluoride (anhydrous) | | 7664-39-3 |
| Hydrogen iodide, anhydrous | | 10034-85-2 |
| Hydrogen selenide | | 7783-07-5 |
| Hydrogen sulfide | | 7783-06-4 |
| Methyl mercaptan | Methanethiol | 74-93-1 |
| Methylchlorosilane | | 993-00-0 |
| Nitric oxide | Nitrogen oxide (NO) | 10102-43-9 |
| Nitrogen trioxide | | 10544-73-7 |
| Nitrosyl chloride | | 2696-92-6 |
| Oxygen difluoride | | 7783-41-7 |
| Perchloryl fluoride | | 7616-94-6 |

*continued*

**TABLE D-3** Continued

| Chemical of Concern | Synonym | CAS Registry Number |
|---|---|---|
| Phosgene | Carbonic dichloride or carbonyldichloride | 75-44-5 |
| Phosphine | | 7803-51-2 |
| Phosphorus trichloride | | 7719-12-2 |
| Selenium hexafluoride | | 7783-79-1 |
| Silicon tetrafluoride | | 7783-61-1 |
| Stibine | | 7803-52-3 |
| Sulfur dioxide (anhydrous) | | 7446-09-5 |
| Sulfur tetrafluoride | Sulfur fluoride ($SF_4$), (T-4)- | 7783-60-0 |
| Tellurium hexafluoride | | 7783-80-4 |
| Titanium tetrachloride | Titanium chloride ($TiCl_4$) (T-4)- | 7550-45-0 |
| Trifluoroacetyl chloride | | 354-32-5 |
| Tungsten hexafluoride | | 7783-82-6 |

**SOURCE:** U.S. Department of Homeland Security list of Chemicals of Interest (6 CFR Part 27 Appendix to Chemical Facility Anti-Terrorism Standards; Final Rule; November 20, 2007).

**TABLE D-4** Examples of Acutely Hazardous Chemicals (Globally Harmonized System Category 1)

| Chemical of Concern | Synonym | CAS Registry Number |
|---|---|---|
| Acrolein | 2-Propenal or acrylaldehyde | 107-02-8 |
| 2-Aminopyridine | | 462-08-8 |
| Arsenic pentafluoride gas | | 784-36-3 |
| Arsine gas | | 7784-42-1 |
| Benzyl chloride | | 100-44-7 |
| Boron trifluoride | Borane, trifluoro | 7637-07-2 |
| Bromine | | 7726-95-6 |
| Chlorine | | 7782-50-5 |
| Chorine dioxide | Chlorine oxide ($ClO_2$) | 10049-04-4 |
| Chlorine trifluoride | | 7790-91-2 |
| Cyanogen chloride | | 506-77-4 |
| Decaborane | | 17702-41-9 |
| Diazomethane | | 334-88-3 |
| Diborane | | 19287-45-7 |
| Dichloroacetylene | | 79-36-7 |
| Dimethylmercury | | 593-74-8 |
| Dimethyl sulfate | | 77-78-1 |
| Dimethyl sulfide | | 75-18-3 |
| Ethylene chlorohydrin | | 107-07-3 |
| Ethylene fluorohydrin | | 371-62-0 |
| Fluorine | | 7681-49-4 |
| 2-Fluoroethanol | | 371-62-0 |
| Hexamethylene diiosocyanate | | 822-06-0 |
| Hydrogen cyanide | Hydrocyanic acid | 74-90-8 |
| Hydrogen fluoride | | 7664-39-3 |
| Iron pentacarbonyl | Iron carbonyl (Fe $(CO)_5$), (Tb5-11)- | 13463-40-6 |
| Isopropyl formate | | 625-55-8 |
| Methacryloyl chloride | | 920-46-7 |
| Methyl acrylonitrile | 2-Propenenitrile, 2-methyl- | 126-98-7 |
| Methyl chloroformate | Carbonochloridic acid, methyl ester | 79-22-1 |
| Methylene biphenyl isocyanate | | 101-68-9 |
| Methyl fluoroacetate | | 453-18-9 |
| Methyl fluorosulfate | | 421-20-5 |
| Methyl hydrazine | Hydrazine, methyl- | 60-34-4 |

*continued*

**TABLE D-4** Continued

| Chemical of Concern | Synonym | CAS Registry Number |
|---|---|---|
| Methyl mercury and other organic forms | | --- |
| Methyl trichlorosilane | | 75-79-6 |
| Methyl vinyl ketone | | 78-94-4 |
| Nickel carbonyl | | 13463-39-3 |
| Nitrogen dioxide | | 10102-44-0 |
| Nitrogen tetroxide | | 10544-72-6 |
| Nitrogen trioxide | | 10544-73-7 |
| Osmium tetroxide | | 20816-12-0 |
| Oxygen difluoride | | 7783-41-7 |
| Pentaborane | | 19624-22-7 |
| Perchloromethlyl mercaptan | Methanesulfenyl chloride, trichloro- | 594-42-3 |
| Phosgene | Carbonic dichloride or carbonyl dichloride | 75-45-5 |
| Phosphine | | 1498-40-4 |
| Phosphorus oxychloride | Phosphoryl chloride | 10025-87-3 |
| Phosphorus pentafluoride | | 7641-19-0 |
| Phosphorus trichloride | | 7719-12-2 |
| Sarin | *O*-Isopropyl methylphosphonofluoridate | 107-44-8 |
| Selenium hexafluoride | | 7783-79-1 |
| Silicon tetrafluoride | | 7783-61-1 |
| Sodium azide | | 26628-22-8 |
| Sodium cyanide (and other cyanide salts) | | 143-33-9 |
| Stibine | | 10025-91-9 |
| Sulfur monochloride | | 10025-67-9 |
| Sulfur pentafluoride | | 10546-01-7 |
| Sulfur tetrafluoride | Sulfur fluoride ($SF_4$), (T-4)- | 7783-60-0 |
| Sulfuryl chloride | | 7791-25-5 |
| Tellurium hexafluoride | | 7783-80-4 |
| Tetramethyl succinonitrile | | 3333-52-6 |
| Tetranitromethane | Methane, tetranitro- | 509-14-8 |
| Thionyl chloride | | 7719-09-7 |
| Toluene-2,4-diisocyanate | | 584-84-9 |
| Trichloro(chlormethyl)silane | | 1558-25-4 |
| Trimethyltin chloride | | 1066-45-1 |

SOURCE: U.S. Department of Homeland Security list of Chemicals of Interest (6 CFR Part 27 Appendix to Chemical Facility Anti-Terrorism Standards; Final Rule; November 20, 2007).

TABLE D-5 Chemicals Used in Clandestine Production of Illicit Drugs

| Chemical of Concern | Target Product |
|---|---|
| Acetic acid | Phenyl-2-propanone (P-2-P)/cocaine |
| Acetic anhydride | Heroin/P-2-P/methaqualone |
| Acetone | Cocaine/heroin/others |
| Acetyl chloride | Heroin |
| *N*-Acetylanthranilic acid | Methaqualone |
| Ammonium formate | Amphetamines |
| Ammonium hydroxide | Cocaine/others |
| Anthranilic acid | Methaqualone |
| Benzaldehyde | Amphetamines |
| Benzene | Cocaine |
| Benzyl chloride | Methamphetamine |
| Benzyl cyanide | Methamphetamine |
| 2-Butanone (methyl ethyl ketone) | Cocaine |
| Butyl acetate | Cocaine |
| *N*-Butyl alcohol | Cocaine |
| Calcium carbonate | Cocaine/others |
| Calcium oxide/hydroxide | Cocaine/others |
| Chloroform | Cocaine/others |
| Cyclohexanone | Phencyclidine (PCP) |
| Diacetone alcohol | Cocaine |
| Diethylamine | Lysergic acid diethylamide (LSD) |
| Ephedrine | Methamphetamine |
| Ergometrine (ergonovine) | LSD |
| Ergotamine | LSD |
| Ethyl acetate | Cocaine |
| Ethyl alcohol | Cocaine/others |
| Ethyl amine | Ethy lamphetamine/3,4-methylenedioxy-*N*-ethylamphetamine (MDE) |
| Ethyl ether | Cocaine/heroin/others |
| *N*-Ethy1ephedrine | Ethylamphetamine/MDE |
| *N*-Ethylpseudoephedrine | Amphetamines |
| Formamide | Cocaine |
| Hexane | Methamphetamine |
| Hydriodic (hydriotic) acid | Cocaine/heroin/others |
| Hydrochloric acid | Cocaine |
| Isopropyl alcohol | Cocaine |
| Isosafrole | Cocaine |
| Kerosene | LSD |
| Lysergic acid | Cocaine |
| Methyl alcohol | Methamphetamine/3,4-methylenedioxymethamphetamine (MDMA) |
| Methylamine | |
| Methylene chloride | Cocaine/heroin/others |
| 3,4-Methylenedioxyphenyl-2-propanone | 3,4-Methylenedioxyamphetamine (MDA)/MDMA/MDE |
| *N*-Methylephedrine | Amphetamines |
| *N*-Methylpseudoephedrine | Amphetamines |

*continued*

**TABLE D-5** Continued

| Chemical of Concern | Target Product |
|---|---|
| Nitroethane | Amphetamines |
| Norpseudoephedrine | 4-Methylaminorex |
| Petroleum ether | Cocaine/others |
| Phenylacetic acid | Phenyl-2-propanone |
| Phenylpropanolamine | Amphetamines/4-methylaminorex |
| 1-Phenyl-2-propanone | Amphetamines/methamphetamine |
| Piperidine | PCP |
| Piperonal | MDA/MDMA/MDE |
| Potassium carbonate | Cocaine |
| Potassium permanganate | Cocaine |
| Propionic anhydride | Fentanyl analogues |
| Pseudoephedrine | Methamphetamine |
| Pyridine | Heroin |
| Safrole | MDA/MDMA/MDE |
| Sodium acetate | P-2-P |
| Sodium bicarbonate | Cocaine/others |
| Sodium carbonate | Cocaine/others |
| Sodium cyanide | PCP |
| Sodium hydroxide | Cocaine/others |
| Sodium suifate | Cocaine/others |
| Sulfuric acid | Cocaine/others |
| Toluene | Cocaine |
| *o*-Toluidine | Methaqualone |
| Xylenes | Cocaine |

**NOTE:** Organizations may opt not to treat the commonly used chemicals on this list as COCs (e.g., acetone).
SOURCE: Severick, James. 1993. *Precursor and Essential Chemicals in Illicit Drug Production: Approaches to Enforcement*, National Institute of Justice. *http://www.popcenter.org/problems/meth_labs/PDFs/Sevick_1993.pdf* (accessed July 2009).

# E

# Sample Safety, Health, and Environment Policy Statement

This university is committed to providing a safe and healthful environment for its employees, students, and visitors and to managing the university in an environmentally sensitive and responsible manner. We further recognize an obligation to demonstrate safety and environmental leadership by maintaining the highest standards and by serving as an example to our students and to the community at large.

The university will strive to improve our safety and environmental performance continuously by adhering to the following policy objectives:

- Developing and improving programs and procedures to ensure compliance with all applicable laws and regulations.
- Ensuring that personnel are properly trained and provided with appropriate safety and emergency equipment.
- Taking appropriate action to minimize hazards or change conditions that endanger health, safety, or the environment.
- Considering safety and environmental factors in all operating decisions, including those related to planning and acquisition.
- Engaging in sound reuse and recycling practices and exploring feasible opportunities to minimize the amount and toxicity of waste generated.
- Using energy efficiently throughout our operations.
- Encouraging personal accountability and emphasizing compliance with standards and conformity with university policies and best practices during employee training and in performance reviews.
- Communicating our desire to improve our performance continu-

ously and fostering the expectation that every employee, student, and contractor on university premises will follow this policy and report any environmental, health, or safety concern to university management.

- Monitoring our progress through periodic evaluations.

Adopted [date] by Safety, Health and Environment Management Committee

# F

# Sample Forms for Chemical Handling and Management

The forms in this appendix are to be used in conjunction with the guidelines for establishing a chemical laboratory safety and security program described in Chapter 3.

## SAMPLE INVENTORY LOGS

**TABLE F-1** Spreadsheet for Laboratory or Building

| Chemical Name and Concentration | Container Type | Location | Responsible Person | Quantity | Allowable Limits |
|---|---|---|---|---|---|
| Hydrogen peroxide, 60% | Glass | 213 Lab Bldg | P. Jones | 4 L | 10 L |
| Dichlorosilane, 100% | Compressed gas | 112 Lab Bldg | R. Solli | 11 lb | 45 lb |

**TABLE F-2** Container Inventory

| Chemical | Location | Responsible Person | Authorized Users |
|---|---|---|---|
| Soman | 215 Lab Bldg | P. Jones | P. Jones<br>L. Martinez<br>K. Liu |
| ***Starting Quantity:*** | *50 g* | ***Date Received:*** | *3 June 2009* |
| **Date** | **Amount Removed** | **Removed by** | **Quantity Remaining** |
| 4 June 2009 | 2 g | K. Liu | 48 grams |
| 5 June 2009 | 4 g | L. Martinez | 44 grams |
| | | | |
| | | | |
| | | | |
| | | | |
| | | | |

# LABORATORY HAZARD ASSESSMENT CHECKLIST

## I. Pre-Operational Planning

| | |
|---|---|
| [ ] Toxicity | What is the level of toxicity? What are the routes of exposure (inhalation, skin absorption, ingestion, injection) and which of these are likely under the conditions of use? What are the signs and symptoms of overexposure? |
| [ ] Flammability | Is the material flammable or explosive under the conditions of use? |
| [ ] Warning Properties | Can odor or irritation adequately warn of over-exposure before it becomes dangerous? |
| [ ] Laboratory Equipment | Is laboratory equipment in good condition? Are machine guards or interlocks in place and functioning? |
| [ ] Storage Precautions | Does the material need isolated storage, refrigeration or other special conditions for storage? |
| [ ] Incompatible Materials | Should certain materials be segregated (e.g., flammables and oxidizers)? |
| [ ] Reagent Stability | Should materials be dated for disposal (e.g., ethers); should materials be kept refrigerated to prolong shelf life? |
| [ ] Protective Clothing | Is a lab apron or clothing made of resistant material needed or is a lab coat adequate? |
| [ ] Gloves | What glove material is needed? Are the right type, thickness, glove length, and size available? |
| [ ] Eye Protection | What type of eye protection is needed—safety glasses for impact, chemical splash goggles for chemicals? Is a face shield needed in combination with the goggles? |
| [ ] Heat Sources | Is heating needed? Is there an alternative to open flames? Are heating mantles in good condition? |
| [ ] Electrical Equipment | Is equipment grounded properly? Are electrical cords insulated? Is ground fault circuit interruption (GFCI) needed? |
| [ ] Vacuum/Pressure Systems | Have connections been leak tested, hydrostatically tested, properly vented, and traps installed when necessary? |
| [ ] Ventilation/Containment | Does the work need to be done in a chemical hood, ventilated cabinet or a glove box to provide the needed level of containment? |

## II. Experimental Scale & Design

| | |
|---|---|
| [ ] Quantity | Are there ways to minimize the amount of materials used without affecting results (e.g., microscale)? |
| [ ] Ambient Conditions | Are special conditions necessary to carry out the reaction (e.g., cold room or dry box)? |
| [ ] Time Constraints | Can the experiment be completed while lab workers are present? If not, can the experiment be safely run unattended or overnight? |

## III. Spill/Emergency Planning

| | |
|---|---|
| [ ] Lab Personnel | Are others in the laboratory aware of what you are doing? |
| [ ] Fire Extinguishers | Are special types required; are you aware of their location and proper use (e.g., Class D for metals)? |
| [ ] Emergency Response | Do you have a response planned in the event of a spill; would evacuation be necessary? |
| [ ] Spill Cleanup | Are materials on hand to absorb/neutralize; is the needed protective equipment on hand and have you been trained on its use? |
| [ ] Safety Shower/ Eyewash Fountain | Are you aware of the locations and methods of operation? |

## IV. Waste Disposal

| | |
|---|---|
| [ ] Method | Is there an approved method for treating the waste in the laboratory? |
| [ ] Labeling | Are containers clearly, indelibly, and accurately labeled as to the contents? |
| [ ] Segregation | Are incompatible wastes kept segregated? |
| [ ] Containers | Are suitable containers with adequate closures available? |
| [ ] Recycling | Is it feasible to safely recover/recycle used chemicals? |

## SAMPLE FORM FOR SAFETY PLAN

### EMERGENCY PREPAREDNESS PLAN FOR WORKING WITH A CHEMICAL

Name________________________ Contact Information ____________________
Building ______________________________________________________
Supervisor ____________________________________________________

### 1. SUBSTANCE INFORMATION

A. Chemical name ______________________________ CAS number ______________
B. Carcinogen  Reproductive toxin  High acute toxicity
C. Estimated rate of use (e.g., g/month) ______________________________
D. MSDS reviewed and readily available  Yes  No

### 2. HAZARDS

#### Physical Hazards

A. Flammable  Yes  No  B. Corrosive  Yes  No
C. Reactive  Yes  No  D. Temperature-sensitive  Yes  No
E. Stability (e.g., decomposes, forms peroxides, polymerizes, shelf-life concerns)  Stable  Unstable
F. Known incompatibilities ____________________________________________

#### Health Hazards

G. Significant routes of exposure
Inhalation hazard  Yes  No
Skin absorption  Yes  No
H. Sensitizer  Yes  No  I. Medical consultation needed  Yes  No

### 3. PROCEDURE

A. Briefly describe how the material will be used ______________________
________________________________________________________________
________________________________________________________________
________________________________________________________________

B. Vacuum system used  Yes  No
C. If yes, describe method for trapping effluents ______________________

## SAMPLE FORM FOR SAFETY PLAN

### 4. EXPOSURE CONTROLS

**Ventilation, Isolation**

A. Chemical hood required Yes No
B. Glove box required Yes No
C. Vented gas cabinet required Yes No

### 5. PERSONAL PROTECTIVE EQUIPMENT (PPE) (Check all that apply)

Safety glasses Chemical-splash goggles Face shield
Gloves (type ____________) Lab coat Apron
Respirator
Other, please describe ____________________________________________

____________________________________________

### 6. LOCATION, DESIGNATED AREA

A. Building ______________________ B. Room ____________
C. Describe below the area where substance(s) will be used ____________

____________________________________________

____________________________________________

D. Location where substances will be stored ____________________
E. Storage method, precautions
Refrigerator/freezer Hood
Special security (describe) Vented cabinet
Flammable liquid storage cabinet Other, describe ________

____________________________________________

### 7. SPILLS AND DECONTAMINATION

A. Spill-control materials readily available Yes No
B. Require special decontamination procedures? If yes, describe. Yes No

____________________________________________

### 8. WASTE DISPOSAL

A. In-lab neutralization Yes No
B. Used up in process (e.g., no waste) Yes No
C. Dispose of as hazardous waste Yes No

# SAMPLE FORM FOR SAFETY PLAN

## 9. AUTHORIZATION

This person has demonstrated an understanding of the hazards of the listed substance and plans to handle the substance in a manner that minimizes risk to health and property. He/she is authorized to use the substance in the manner described.

______________________________ ______________________________

Supervisor Chemical Safety and Security Officer

## 10. USE RECORD (to be completed after the material is used)

A. Describe how the material was disposed of:

B. Explain differences between initial planning and how material was actually used or handled.

C. Are less hazardous materials available to produce the same or better results? If so, describe.

D. Could the quantity or concentration used be reduced for safer handling without causing an unwanted outcome? If so, describe.

E. List any recommendations for improving the health, safety or environmental impact of this process or chemical in the future.

## SAMPLE LABORATORY EMERGENCY INFORMATION SHEET

# LABORATORY EMERGENCY INFORMATION SHEET

| Department | Room | Date |
|---|---|---|
| Manager Responsible for Lab | Office Phone | Home Phone |
| Alternate Contact | Office Phone | Home Phone |
| Alternate Contact | Office Phone | Home Phone |
| Emergency Coordinator:<br>Building or Dept/School | Office Phone | Home Phone |

**IN CASE OF EMERGENCY**, tell your laboratory manager and call ____________.
For **fire**, pull alarm; evacuate building; stay outside to meet with fire department official.
For **hazardous vapors or gases**, inform others to evacuate the area; close doors; call ____________.
For **gases or vapors spreading to other areas**, pull fire alarm; evacuate the building; WHEN IN DOUBT, GET OUT.
For **injuries**, call ______________ for ambulance.
For **poison** and other chemical toxicity information, call ___________.
For **simple spills**, call ____________ for cleanup advice.

| **Institutional Emergency Coordinator:** | **Ambulance/Fire/Police/Spill:** |
|---|---|
| **Hospital Emergency Room:** | **Poison Control Center:** |
| **LOCATION** | **LOCATION** |
| Nearest Fire Extinguisher: | Nearest Fire Alarm: |
| Nearest Spill Control Material: | Nearest Safety Shower: |

| BIOHAZARDS | LAB LOCATION | CHEMICALS | LAB LOCATION | RADIATION | LAB LOCATION |
|---|---|---|---|---|---|
| Biosafety Level 1 ☐ *Low* | | ☐ Flammable Liquid | | ☐ Laser | |
| Biosafety Level 2 ☐ | | ☐ Air/Water Reactive | | ☐ Irradiator | |
| Biosafety Level 3 ☐ | | ☐ Toxics/ Carcinogens | | ☐ Rad. Sealed Source | |
| Biosafety Level 4 ☐ *High* | | ☐ Conc. Acids/Bases | | ☐ Radioactive Waste | |
| Pathogens: | | ☐ Gas Cylinders | | ☐ Rad. Materials | |
| ☐ Human | | ☐ Strong Oxidizers | | ☐ Other: | |
| ☐ Animal | | ☐ Waste Solvents | | | |
| ☐ Toxins | | ☐ Other: | | | |
| ☐ Other: | | ☐ Other: | | | |

Complete and post next to your laboratory door, with a second copy next to your phone.

# G

# Compliance Forms

# SAMPLE INSPECTION CHECKLIST

## INSPECTION CHECKLIST

Department, Group, or Laboratory: ____________________
Inspector: ____________________
Date: ____________________
Building and room: ____________________
Laboratory supervisor: ____________________

### LABORATORY ENVIRONMENT

| | |
|---|---|
| Work areas illuminated | Y N NA |
| Storage of combustible materials minimized | Y N NA |
| Aisles and passageways clear and unobstructed | Y N NA |
| Trash removed promptly | Y N NA |
| No evidence of food or drink in active laboratory areas | Y N NA |
| Wet surfaces covered with nonslip materials | Y N NA |
| Exits illuminated and unobstructed | Y N NA |

COMMENTS:

Other elements that the checklist can include

### EMERGENCY EQUIPMENT AND PLANNING

| | |
|---|---|
| Fire extinguishers mounted and unobstructed | Y N NA |
| Fire extinguishers fully charged with tamper indicators in place | Y N NA |
| Fire extinguisher inspection up to date | Y N NA |
| Eyewash unit and safety shower within 10 seconds of hazard | Y N NA |
| Eyewash unit and safety shower inspection up to date | Y N NA |
| Fire alarm pull stations unobstructed | Y N NA |
| Spill control materials available and adequate for potential spills | Y N NA |

COMMENTS:

# SAMPLE INSPECTION CHECKLIST

## PERSONAL PROTECTIVE EQUIPMENT

| | |
|---|---|
| Personnel wearing appropriate eye and face protection | Y N NA |
| Personnel wearing appropriate gloves | Y N NA |
| Shoes appropriate to the hazard | Y N NA |
| Clothing appropriate to the hazards posed in the laboratory | Y N NA |

COMMENTS:

## SIGNS, LABELS, PLANS, AND POSTINGS

| | |
|---|---|
| Emergency action plan available | Y N NA |
| Material-safety data sheets accessible | Y N NA |
| Chemical-hygiene plan available | Y N NA |
| Contact sheet posted and up to date | Y N NA |
| Telephones labeled with emergency number | Y N NA |
| Building evacuation routes posted | Y N NA |
| Ice-making machines labeled "Not for human consumption" | Y N NA |
| Chemical refrigerators labeled "No food" | Y N NA |
| Food refrigerators labeled "Food only—no chemicals" | Y N NA |
| Lasers properly labeled | Y N NA |
| High-voltage equipment properly labeled | Y N NA |
| Emergency equipment labeled with highly visible signs | Y N NA |

COMMENTS:

## ELECTRICAL HAZARDS

| | |
|---|---|
| Flexible cords in good condition | Y N NA |
| Cords not on surfaces where flammable liquids may pool | Y N NA |
| Cover plates in place for outlets and switches | Y N NA |
| Circuit-breaker panels unobstructed | Y N NA |
| Multiplug adapters have overload protection | Y N NA |
| No extension cords in use | Y N NA |
| Ground-fault circuit interrupters (GFCI) used for wet areas | Y N NA |
| Guards or covers in place for electrophoresis devices | Y N NA |

COMMENTS:

# SAMPLE INSPECTION CHECKLIST

## STORAGE

|  | | | |
|---|---|---|---|
| Heavy items on lower shelves | Y | N | NA |
| Storage at least 18 in. below sprinkler heads | Y | N | NA |
| Storage at least 24 in. below ceiling | Y | N | NA |
| Means available to reach items stored above shoulder level | Y | N | NA |
| Shelving adequate for loads imposed | Y | N | NA |
| Chemicals stored by compatibility and hazard class | Y | N | NA |
| Chemical containers clearly labeled with contents | Y | N | NA |
| Corrosive chemical stored below eye level | Y | N | NA |
| Materials with shelf-lives dated on receipt | Y | N | NA |
| Secondary containment used near sinks and drains | Y | N | NA |
| Waste containers sealed except during transfers | Y | N | NA |
| Waste containers labeled with contents, "Hazardous Waste" | Y | N | NA |
| Storage limited to less than 1 quart of acutely hazardous waste | Y | N | NA |
| Storage limited to less than 55 gallons hazardous waste | Y | N | NA |

COMMENTS:

## COMPRESSED GASES AND CRYOGENICS

|  | | | |
|---|---|---|---|
| Toxic, flammable, corrosive gases used in chemical-fume hood | Y | N | NA |
| Stored upright, secured from tipping | Y | N | NA |
| Regulator compatible with gas cylinder | Y | N | NA |
| Cylinder carts used for transport | Y | N | NA |
| Valve caps in place when not in use | Y | N | NA |
| Empty or unused cylinders returned to supplier | Y | N | NA |
| Gases and cryogenic liquids dispensed with good ventilation | Y | N | NA |
| Cryogenic dewars vented or have pressure-relief devices | Y | N | NA |
| Glass dewars shielded | Y | N | NA |

COMMENTS:

## PRESSURE AND VACUUM SYSTEMS

|  | | | |
|---|---|---|---|
| Vacuum glassware in good condition | Y | N | NA |
| Vacuum pressure-relief devices in place and inspected | Y | N | NA |
| Glass vessels shielded or enclosed | Y | N | NA |
| Temperature and pressure measuring devices in place where needed | Y | N | NA |

COMMENTS:

# SAMPLE INSPECTION CHECKLIST

## CHEMICAL HOODS AND VENTILATION

| | | | |
|---|---|---|---|
| Each chemical fume hood tested within last year | Y | N | NA |
| Sash closed when not in active use | Y | N | NA |
| Chemical-fume hood vents (baffles) unobstructed | Y | N | NA |
| Chemical-fume hood used with sash in appropriate position | Y | N | NA |
| Chemical storage limited in actively used hood | Y | N | NA |
| Chemicals and equipment at least 6 in. from the sash | Y | N | NA |

COMMENTS:

## SECURITY

| | | | |
|---|---|---|---|
| Doors to lab operate, close and lock properly | Y | N | NA |
| Windows operate, close, and lock properly | Y | N | NA |
| Alarm systems operating properly | Y | N | NA |
| Keys and access cards kept in secure area out of sight | Y | N | NA |

COMMENTS:

## TRAINING AND AWARENESS

| | | | |
|---|---|---|---|
| Workers have attended all appropriate training | Y | N | NA |
| Training has been documented | Y | N | NA |
| Laboratory personnel know... | | | |
| What to do in event of an emergency, such as fire or injury | Y | N | NA |
| How to clean up chemical spills | Y | N | NA |
| Location and contents of the chemical hygiene plan | Y | N | NA |
| Chemical hygiene officer or safety manager | Y | N | NA |
| What MSDSs are and where to find them and other safety info | Y | N | NA |
| What type of personal protective equipment to use and when to use it | Y | N | NA |
| What to do with chemical waste | Y | N | NA |
| What are the most hazardous materials they use and what precautions to take | Y | N | NA |
| Where and how to use safety showers and eyewash units | Y | N | NA |
| To question unfamiliar visitors in the lab | Y | N | NA |
| How and when to report injuries, illnesses, or incidents | Y | N | NA |

COMMENTS:

# INCIDENT REPORT FORM

## INCIDENT REPORT

**PERSONAL DATA**

| | | | |
|---|---|---|---|
| Employee/Student Name | | | Case No. |
| | | Employee/Student Phone No. | |
| Employee/Student Dept. | | Investigation Date | |
| Employee Supervisor | | Investigator Name | |

**EVENTS DETAILS**

**Employee/Student Statement** (Description of event—before, during, and after)

______________________________________________

______________________________________________

______________________________________________

______________________________________________

______________________________________________

______________________________________________

| | | | |
|---|---|---|---|
| Work Related? | Yes No | Body Part Injured | |
| Event Date/Time | / | Event Location | (lab, corridor, stairs, outside, etc.) |
| Reported Inquiry Date/Time | / | Specific Location | (building, floor, room, column) |

| | | |
|---|---|---|
| Injury Severity | Observation/Near Miss | First Aid |
| | Work Restrictions | Lost Time Restrictions |

| | | | |
|---|---|---|---|
| **Accident Type** | Allergen Exposure | Bitten By | |
| | Car/Truck/Motorized Vehicle | | |
| | Caught In/Between | Contact with Chemical | Contact with Hot Surface |
| | Environmental Exposure | Ergonomic | Needle Stick |
| | Pushing/Pulling | Slip/Trip/Fall | Struck Against |
| | Struck By | Twist/Turn | Other |

| | **Device Type** | **Device Brand** |
|---|---|---|
| Contaminated Sharp Involved | | |
| Needle Stick | | |

# INCIDENT REPORT FORM

Allergic Agent ______________________________

Chemicals or Biohazards Involved______________________________

Equipment Involved / ID Number______________________________

## DESCRIBE POSSIBLE CAUSES

Equipment ______________________________
Tools / PPE______________________________
Environment ______________________________
Procedure______________________________
Personnel______________________________
Other ______________________________

## CAUSAL FACTORS

______________________________
______________________________
______________________________
______________________________
______________________________
______________________________
______________________________

## RECOMMENDATIONS

| Corrective Actions/Preventative Actions | Person Responsible | Due Date |
|---|---|---|
| | | |
| | | |
| | | |
| | | |
| | | |

## EXAMPLE CLASSROOM LESSON FOR LABORATORY MANAGERS

### Instructors' Guide

In this Instructors' Guide, there are 10 lessons [one example is provided here] to be used in training laboratory staff, students, and volunteers. Each lesson contains the following elements:

- An introduction that summarizes the content of the lesson and provides ideas for teaching the content to training participants;
- Objectives, or concepts that all participants should master after studying each lesson;
- One or more segments that describe a problematic situation in a laboratory;
- Questions for participants to answer and discuss as a group; and
- Commentary on each question for Instructors to use in guiding participants in their discussions.

The text and questions of each lesson should be handed out to participants if possible. The introduction and commentary for each lesson are for use by the instructor *only* and should *not* be distributed to participants. The introduction and commentary are available to help the instructor guide the discussion, ask the appropriate questions, and make the experience useful for all participants.

Lessons provide an effective method of teaching. Discussing lessons is a way to involve participants in familiar and relevant issues. The purpose of the lessons is to ask participants to consider the choices they face as they attempt to promote a culture of safety and security in the laboratory.

Five of the lessons are directed toward laboratory managers or others who supervise laboratories. The remaining four lessons are directed toward people who work in the laboratories, including students and employees. Discussion begins with participants thinking about what might be going on in the minds of the fictional individuals featured in the cases. Every lesson includes reflective questions that aim to encourage participants to consider the following concerns: *Could this happen in our laboratory? Does this happen here? What strategies could we develop to deal with this issue in our workplace?*

Below are tips on how to successfully use lessons. The below guidance is taken from Kenneth D. Pimple's article "Using Case Studies in Teaching Research Ethics."[1] Pimple says that you must:

[1]Adapted from Pimple, Kenneth D., "Using Case Studies in Teaching Research Ethics" (2007). Resources. Paper 293. http://www.ethicslibrary.org/resources/293.

- prepare in advance to lead the lesson discussion. Decide what goals to accomplish, how to discuss the situations presented, and how much time to spend on each lesson.
- set ground rules at the beginning of the session. Remind participants to be open, honest, and respectful.
- offer participants broad strategies and tactics before discussing lessons. Some of these tactics include:
  - thinking about immediate, near-future, and long-term steps to take;
  - thinking about what might be going on in the minds of the fictional people featured in the lesson;
  - considering strategies to deal with the problem in the laboratory; and
  - taking a personal role in the problem—*What would I do in this situation?*

For each lesson discussion, follow these suggested procedures:

- Before starting each lesson discussion, distribute copies of the lesson to participants to make it easier for them to participate.
- Ask one of the participants to read the lesson aloud. This allows participants to be engaged at an early stage.
- Give participants about five minutes to think about the lesson individually, write down any thoughts they may have, and answer the questions before discussing them aloud.
- After participants have been given time to work independently, have them share short responses to the lesson. Then allow participants to discuss answers to the questions.
- As instructor, listen to the discussion without actively participating, unless the discussion becomes disorderly or off-point.

The goal of instructors is to build trust and encourage honest reflection. Encourage participants to work independently or as a group to devise concrete strategies for dealing with the issues presented in the lessons. Strategies should include immediate steps and future steps.

At the end of each lesson, participants should recognize some of the barriers that prevent laboratory personnel from behaving in a safe or secure manner and should be able to list steps for overcoming those barriers. Only by addressing barriers can laboratory personnel promote a culture of safety and security. In addition, participants should leave the training session feeling empowered to think creatively in response to safety issues. Finally, it is important for all participants to leave the session understanding that everyone is responsible for the safety and security of the laboratory, not just certain individuals.

## Lesson 1: Ensuring the Use of Safety Measures in the Laboratory

### For Instructor

**Overview:** This lesson describes the challenges a new laboratory manager faces in ensuring that staff uses appropriate personal protective equipment.

**Objectives:**

- Recognize the importance of personal protective equipment (PPE)
- Identify barriers to the safe and consistent use of PPE in laboratories
- Generate action steps that laboratory managers can use to encourage the use of PPE among laboratory workers and visitors
- Identify methods for convincing supervisors and other institutional leaders of the importance of PPE and its regular use
- Recognize that there are many ways to encourage the adoption of safe practices in a laboratory

**Reasons for not wearing the goggles in this lesson could include**

- cost and/or unavailability of goggles;
- a lack of habit;
- a lack of initial understanding of the importance of wearing protective equipment or of the hazards posed by the work;
- a sense of invincibility;
- a lack of confidence or respect in the new laboratory manager;
- a cultural acceptance of risk and destiny;
- feeling of resistance since the new laboratory manager is an outsider;
- workers wanting to rebel against the changes to traditional practices;
- possible anticipation of physical discomfort from wearing the goggles;
- laboratory manager's supervisor's attitude; and
- feeling of unattractiveness or detracting from their physical appearance.

There may also be many reasons why the women in the lab are even less likely to use the goggles than the men. Perhaps the female workers believe their tasks are less risky than the ones performed by the men. It is also possible that the women may feel that their health is less important than that of the men in their laboratories and if so, perhaps they are choosing not to diminish the supply of available goggles for their male colleagues.

**Segment 1**

A recent graduate of a well-respected institution is hired as a laboratory manager for a small chemical company. Soon after starting work, the manager notices that many laboratory personnel do not have safety goggles. To fix the problem, the manager orders pairs for everyone and invites the staff to pick them up from central inventory. A few weeks later, the manager notices that many pairs of goggles are still in storage. On a walk through the labs to see what is going on, the manager notices that many of the goggles are prominently displayed on laboratory shelves but still in boxes. The manager also notices that many of the female employees have not even picked up their goggles from central inventory.

1. **Why would the laboratory personnel be reluctant to wear the safety goggles?**
   *Instructor: Encourage participants to share why there is a disregard for safety and consider what could be influencing the workers' actions. Please refer to the previous page for examples of why personnel may not be wearing goggles.*
2. **What should the lab manager do?**
   *Instructor: Lead a discussion to find the best course of action. Participants' suggestions could include:*
   - *hold a training session for all laboratory personnel that focuses on the importance of PPE and its regular use;*
   - *post signs in the laboratory to remind personnel of the importance of PPE, especially goggles, and its correct use;*
   - *regularly remind personnel to use goggles, and check on their proper use;*
   - *distribute uncollected goggles to personnel; and*
   - *praise and thank people for properly wearing PPE.*

**Segment 2**

As an attempt to rectify the situation, the laboratory manager hands out the remaining goggles to those who had neglected to pick them up and reminds the staff of the importance of using safety goggles while working in the lab. The manager is reassured by the fact that everyone agreed with him. Nevertheless, when walking through the labs a few days later, the manager notes again that many personnel still are not wearing their goggles.

3. **What should the lab manager do now? List the strategies the manager could use in the table below. Note the advantages and disadvantages of each option.**

*Instructor: Encourage participants to think about immediate, near-term, and long-term strategies. An example is provided in the first row of the table.*

| ***Barriers*** | ***Strategies to address barrier*** | ***Advantages*** | ***Disadvantages*** |
|---|---|---|---|
| *goggles are uncomfortable to wear* | *remind personnel of the need for goggles to prevent eye injuries* | *demonstrates a consistent culture of safety in the lab* | *does not address the specific problem of discomfort* |

4. **What kind of help would the manager need? From whom?**
   *Instructor: Help participants recognize that it is much harder to make change alone. Some solutions could include seeking help from peers, supervisors, or professional societies.*
5. **Would the situation be different if the laboratory manager were an older, established researcher?**
   *Instructor: Encourage participants to share what they would do if the laboratory manager were an established, older researcher. Ask questions to help participants understand the difficulties faced by younger managers, such as: Do workers listen to older, more experienced managers more than someone who is younger and recently graduated?*
6. **If the laboratory manager were a woman, would the situation be different? How?**
   *Instructor: Have participants discuss the impacts of having a female laboratory manager in their own laboratory. Encourage participants to recognize any special challenges that a female manager might have that a male manager might not face. Lead participants to brainstorm ways that a female manager might overcome these challenges.*

**Segment 3**

After many weeks of work, the manager succeeds in getting the staff to wear their goggles consistently. One day, as part of a review of the institution, the manager's supervisor takes a tour of the laboratories. When offered goggles before entering the laboratory, the supervisor waves them off saying, "Oh, I will only be in there for a few minutes. I'm sure I'll be fine."

1. **What impact could the supervisor's behavior have on the laboratory staff?**
   *Instructor: Help participants recognize that the supervisor acts as a role model. Not wearing goggles may undo the work put in by the laboratory manager.*
2. **What should the laboratory manager do now?**
   *Instructor: Ask participants to discuss the position the laboratory manager is in. Lead the discussion with questions such as: What is going through the manager's head? Should the manager publicly challenge the supervisor, or is it best for the manager to talk to the supervisor in private?*

   *Have participants write down all of the options available to the laboratory manager and discuss them to find the best answer. Suggestions could include discussing the situation in private with the supervisor, trying to gently encourage the supervisor to wear the goggles, pointing out that the entire laboratory staff wears goggles regularly, or remind the supervisor that it is really important to set an example for others who look up to him or her. Remind participants that the manager's goal is to have the supervisor support the manager and the culture of safety throughout their institution.*
3. **How is this case relevant to your laboratory?**
   *Instructor: Ask participants to draw from their own experiences. Ask questions such as: Have you ever been in a situation similar to this? What did you do?*
4. **Are safety goggles required in your laboratory at all times? Why or why not?**
   *Instructor: This is a reflective question for participants. Encourage participants to think about reasons why they don't wear goggles. Ask: Are there times when it is not necessary to wear goggles? Then discuss how to address this issue.*
5. **Does the staff in your laboratory comply with other similarly important safety measures? Why or why not?**
   *Instructor: This question should be used for self-reflection. Ask participants to write down answers for themselves only, as a way to reflect on their performance as managers.*
6. **What strategies should laboratories put in place to better promote a culture of safety?**
   *Instructor: Write down participants' suggestions on a board or large piece of paper. Ideas could include training, signage, better leadership compliance, or investment in PPE and other safety equipment.*

7. **To better promote a culture of safety, what support will laboratory managers need? From whom?**
   *Instructor: Ask participants to think about what resources they will need to successfully implement the strategies suggested in number 6 above. Encourage participants to seek help from supervisors, peers, other institutions, and professional societies.*
8. **What is the best way to secure that support for a culture of safety?**
   *Instructor: The answer to this question will be dependent on each individual institution. Some countries may not have ample resources to implement all safety strategies.*

## For Participants

### Segment 1

A recent graduate of a well-respected institution is hired as a laboratory manager for a small chemical company. Soon after starting work, the manager notices that many laboratory personnel do not have safety goggles. To fix the problem, the manager orders pairs for everyone and invites the staff to pick them up from central inventory. A few weeks later, the manager notices that many pairs of goggles are still in storage. On a walk through the labs to see what is going on, the manager notices that many of the goggles are prominently displayed on laboratory shelves but still in boxes. The manager also notices that many of the female employees have not even picked up their goggles from central inventory.

1. **Why would the laboratory personnel be reluctant to wear the safety goggles?**

2. **What should the lab manager do?**

### Segment 2

As an attempt to rectify the situation, the laboratory manager hands out the remaining goggles to those who had neglected to pick them up and reminds the staff of the importance of using safety goggles while working in the lab. The manager is reassured by the fact that everyone agreed with him. Nevertheless, when walking through the labs a few days later, the manager notes again that many personnel still are not wearing their goggles.

1. **What should the lab manager do now? In the table below, list the strategies the manager could use. Note the advantages and disadvantages of each option.**

| *Barriers* | *Strategies to address barrier* | *Advantages* | *Disadvantages* |
|---|---|---|---|
| | | | |

2. **What kind of help would the manager need? From whom?**

3. **Would the situation be different if the laboratory manager were an older, established researcher?**

4. **If the laboratory manager were a woman, would the situation be different? How?**

Segment 3

After many weeks of work, the manager succeeds in getting the staff to wear their goggles consistently. One day, as part of a review of the institution, the manager's supervisor takes a tour of the laboratories. When offered goggles before entering the laboratory, the supervisor waves them off saying, "Oh, I will only be in there for a few minutes. I'm sure I'll be fine."

1. **What impact could the supervisor's behavior have on the laboratory staff?**

2. **What should the laboratory manager do now?**

3. **How is this case relevant to your laboratory?**

4. **Are safety goggles required in your laboratory at all times? Why or why not?**

5. **Does the staff at your laboratory comply with other similar important safety measures? Why or why not?**

6. **What strategies should laboratories put in place to better promote a culture of safety?**

7. **To better promote a culture of safety, what support will laboratory managers need? From whom?**

8. **What is the best way to secure that support for a culture of safety?**